Vorwort zur Patentsammlung Puppenhäuser

Seit dem frühen Mittelalter begleiten Musterbücher und Musterblätter große Meister und ihre Familien. In ihnen wurde das über Generationen gesammelte Wissen bewahrt und an die Nachfolger und Erben weitergegeben. Patentschriften sind moderne Musterbücher für Modellbauer. Patente sind die jeweils weltweit erste Dokumentation einer neuen Idee und belegen auf besondere Art das Fortschreiten der Technik und Ihre Geschichte. Aus der riesigen internationalen Sammlung technischer Lösungen mit über 200 einzigartigen innovativen Ideen zum Thema Puppenhaus wurden die vorliegenden repräsentativen Patentschriften ausgewählt und zu einem besonderen Musterbuch zusammengeführt. Jede Schrift enthält eine umfassende technische Beschreibung der Erfindung und detailgenaue Zeichnungen. Konzept, Konstruktion und Funktion lassen sich anhand der Darstellung in allen Details nachvollziehen.

Ungefähr 90 % des weltweiten technischen Wissens ist in Patentschriften veröffentlicht. Von diesen Patenten sind jedoch nur ca. 10% in Kraft. Bei über 90 % der registrierten Erfindungen ist der Rechtsschutz inzwischen verfallen und die Nutzung kostenlos möglich. Alle Puppenhaus-Patente, die in dieser Sammlung zusammengetragen wurden, sind mit Erscheinen dieses Buches frei von Patentrechten. Die technischen Problemlösungen sind somit frei von Schutz-Rechten und können jederzeit privat genutzt werden. Die gewerbliche Verwendung kann trotzdem weiterhin verwandten Schutzrechten wie z.B.: dem unlauteren Wettbewerb UWG unterliegen und somit eingeschränkt sein.

Zeichnen Sie Ihre eigenen Bilder, entwerfen Sie Ihr eigenes Werk,
und werden Sie selbst ein Schöpfer!
Denn die IDEE ist frei und unterliegt keinem Schutzrecht!

St. Helena, 07. August 2014   K. Winter

Bibliographische Information der Deutschen Nationalbibliothek
Die Deutsche Nationalbibliothek verzeichnet diese Publikation in der Deutschen Nationalbibliografie;
detaillierte bibliografische Daten sind im Internet über http://dnb.d-nb.de abrufbar.

Reihe: Vorlage - Konzept - Konstruktion
Band 2, 1. Auflage 2014

Lektorat sowie Entwurf und Gestaltung,
Arrangement der Texte, Bilder und Grafiken
durch ATELIER-KALAI

www.ATELIER-KALAI.de

Herstellung und Verlag
Books on Demand GmbH, Norderstedt
ISBN 978-3-7357-9424-6

INHALTSANGABE
Patentsammlung Puppenhäuser II
System und Technik

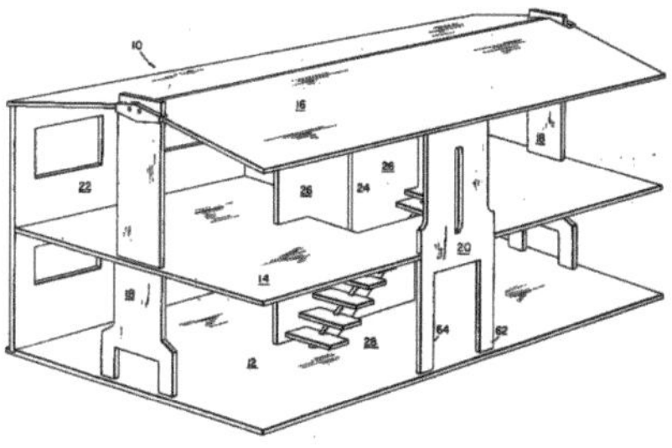

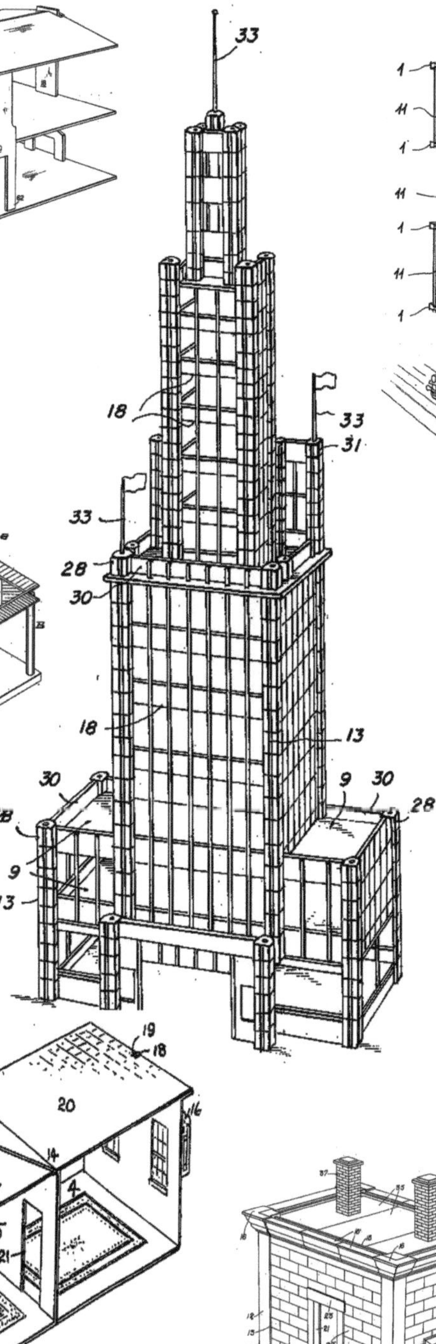

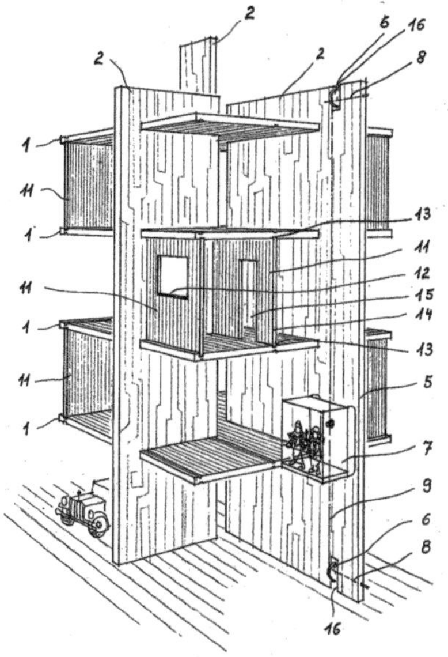

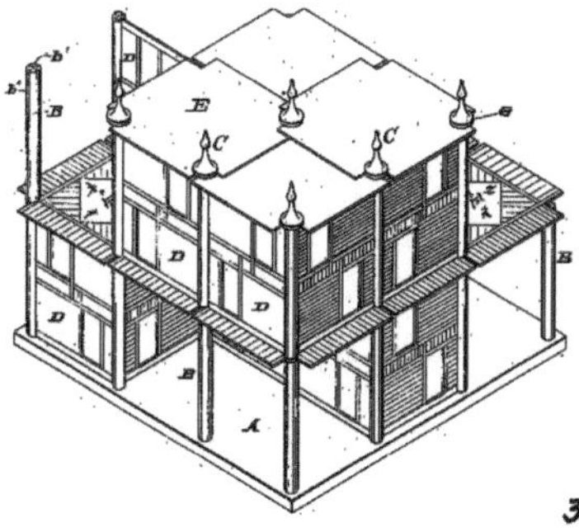

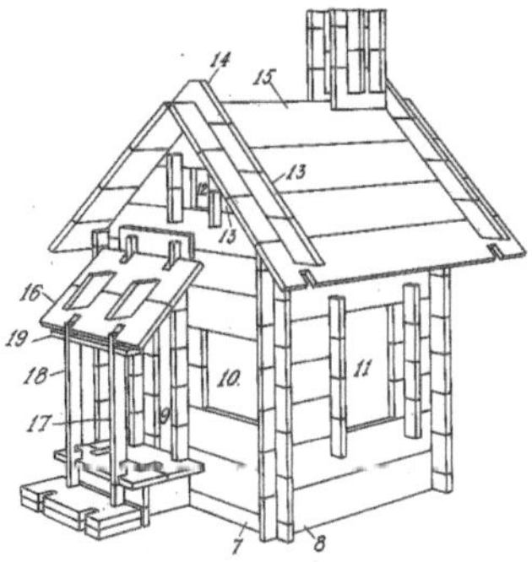

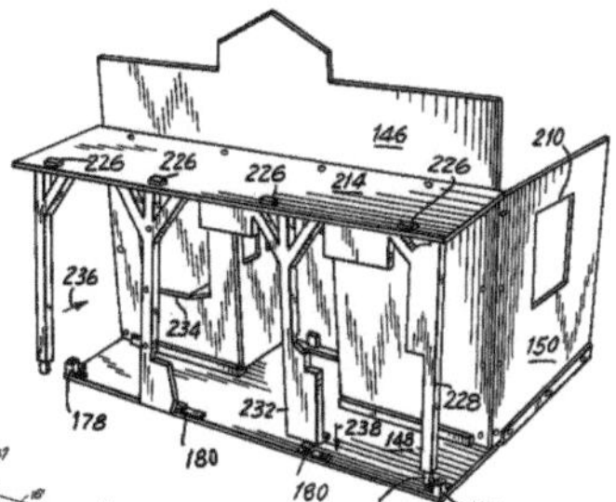

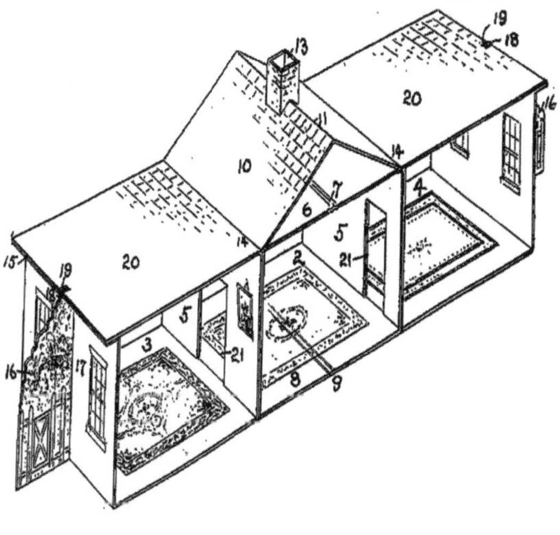

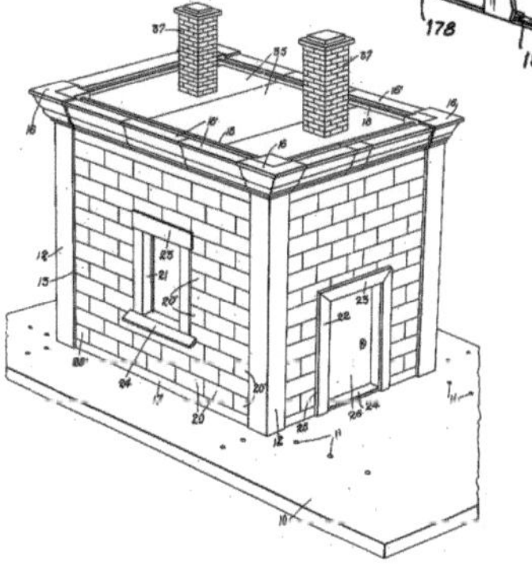

Patentsammlung Puppenhäuser II • System und Technik • Reihe: Vorlage - Konzept - Konstruktion • Band: 2
Herausgeber: www.atelier-kalai.de, Kerstin Winter • Hersteller / Verlag: Books on Demand GmbH Norderstedt

Bisher bei Atelier–Kalai erschienen:

• Kunstschmieden / Ironwork / Ferronnerie / Hierro Forjado I
Gitter und Geländer aus Schmiedeeisen
(sehen-wissen-gestalten Band 1)

• Musterbuch Renaissance Stickereien
Stickvorlagen für Blumen Tiere und Waldfrüchte
(sehen-wissen-gestalten Band 2, 2. Aufl)

• Musterbuch Blumen sticken
(sehen-wissen-gestalten Band 3)

• Chinesische Gartenhäuser und Pavillons
(sehen-wissen-gestalten Band 4)

• Patentsammlung Puppenhäuser I
(Vorlage-Konzept-Konstruktion Band 1)

• Patentsammlung Puppenhäuser II
(Vorlage-Konzept-Konstruktion Band 2)

Patentsammlung Puppenhäuser II • System und Technik • Reihe: Vorlage - Konzept - Konstruktion • Band: 2
Herausgeber: www.atelier-kalai.de, Kerstin Winter • Hersteller / Verlag: Books on Demand GmbH Norderstedt

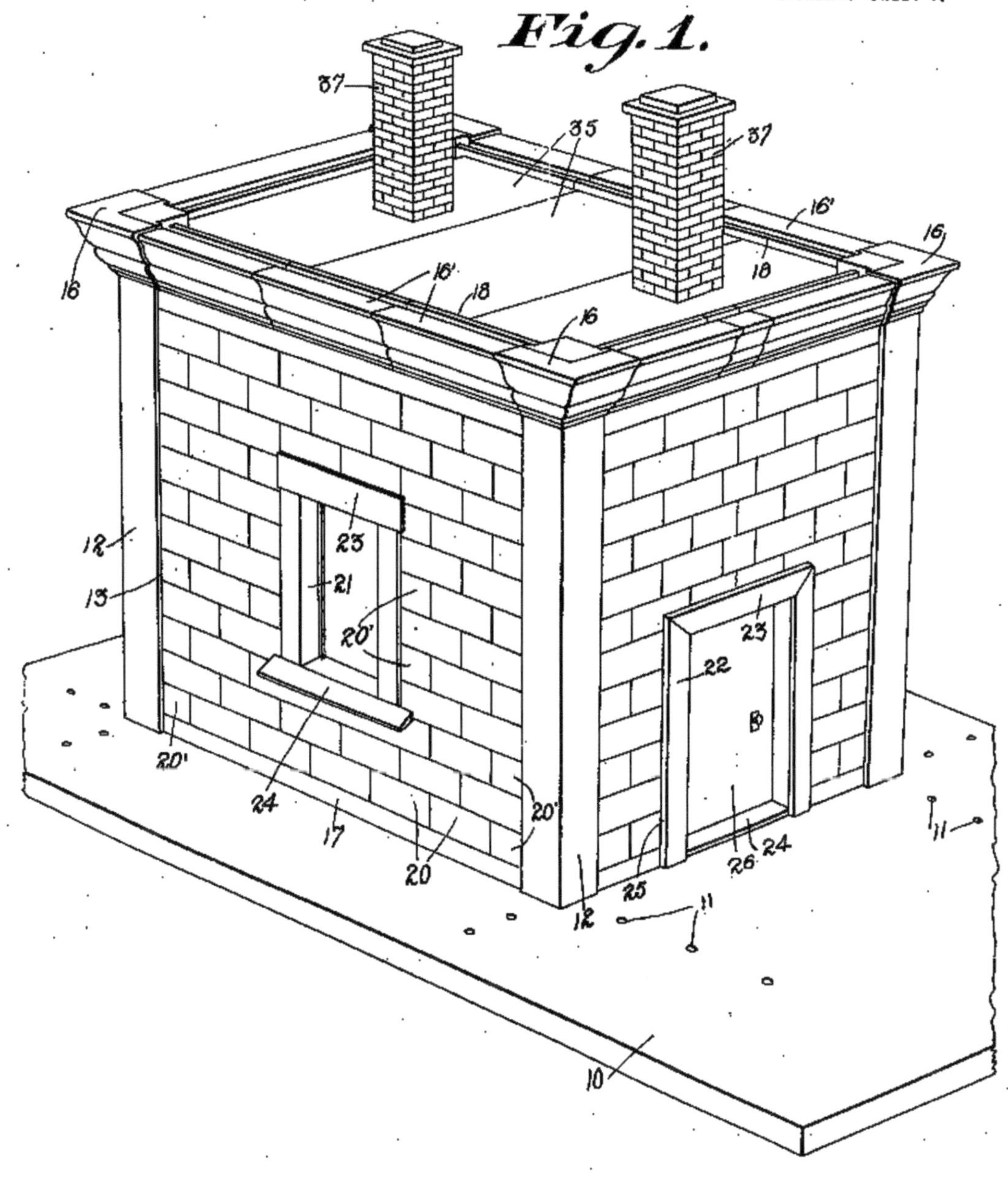
O. W. WARD.
BLOCK TOY BUILDING.
APPLICATION FILED JUNE 18, 1919.
1,337,171.
Patented Apr. 13, 1920.
3 SHEETS—SHEET 1.
Fig.1.
WITNESSES
INVENTOR
O. W. Ward
BY
ATTORNEYS

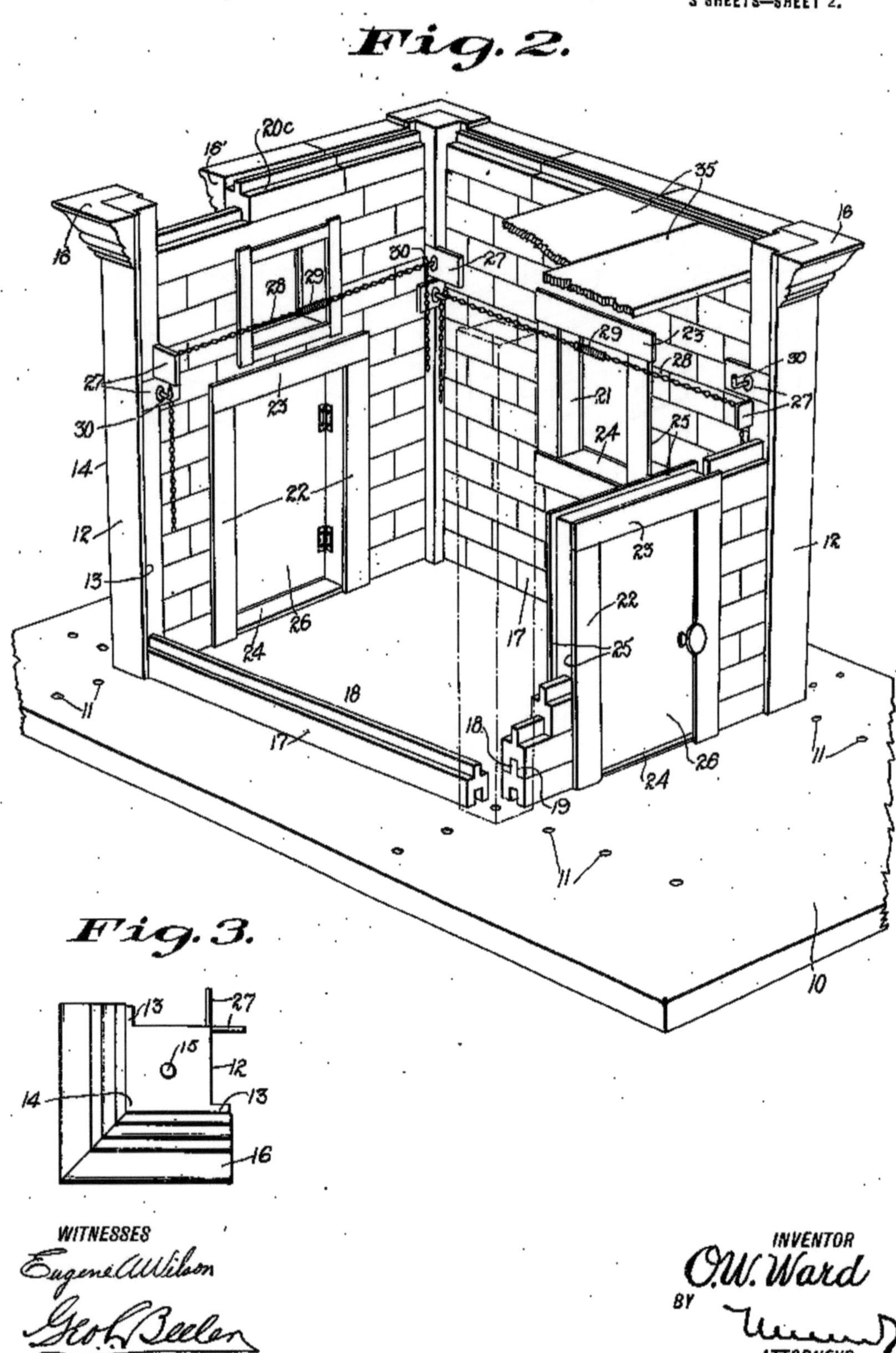

O. W. WARD.
BLOCK TOY BUILDING.
APPLICATION FILED JUNE 18, 1919.
1,337,171.
Patented Apr. 13, 1920.
3 SHEETS—SHEET 2.
Fig. 2.
Fig. 3.
WITNESSES
Eugene A. Wilson
Geo. C. Beeler
INVENTOR
O. W. Ward
BY
ATTORNEYS

O. W. WARD.
BLOCK TOY BUILDING.
APPLICATION FILED JUNE 18, 1919.

1,337,171.

Patented Apr. 13, 1920.
3 SHEETS—SHEET 3.

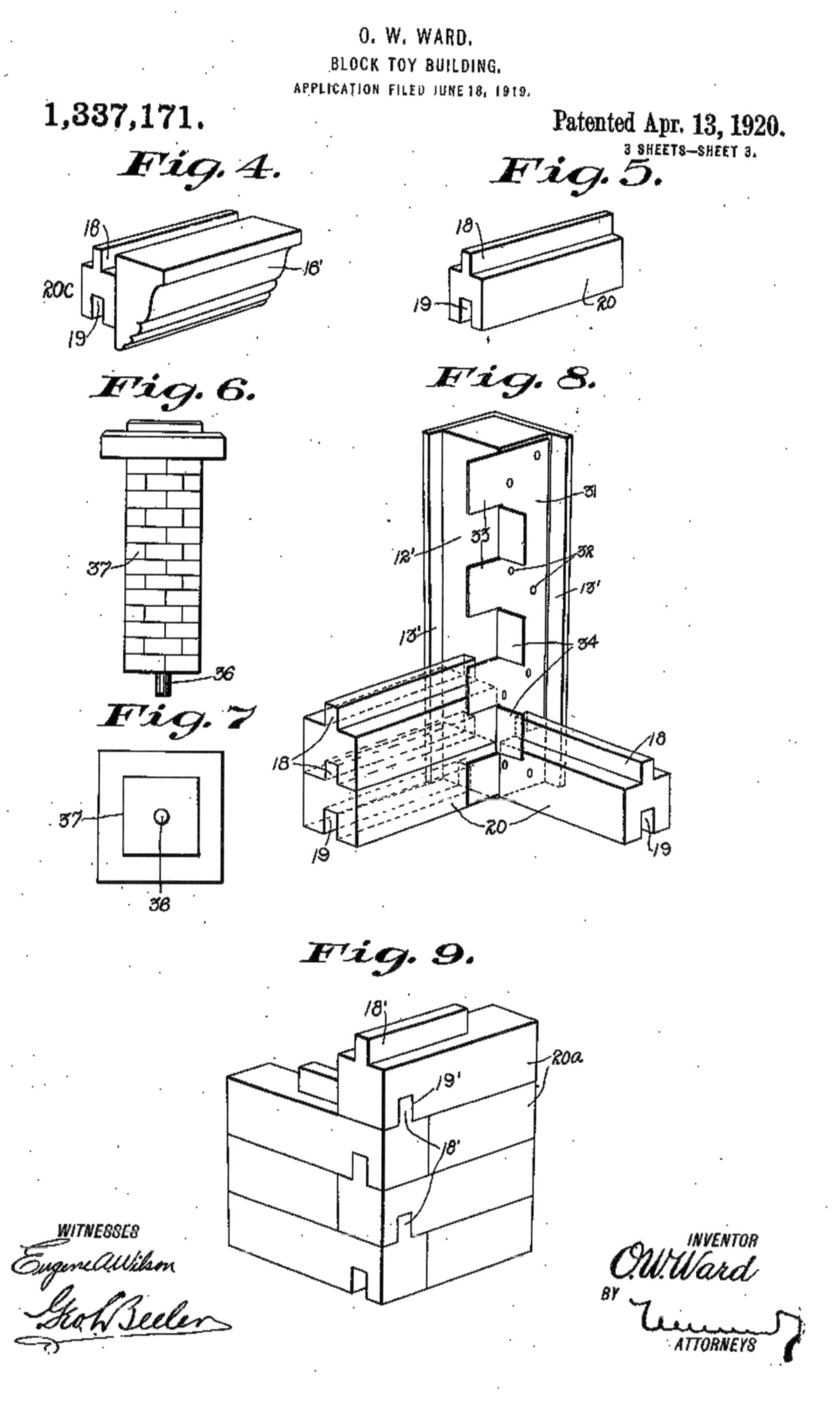

UNITED STATES PATENT OFFICE.

ORIN WARREN WARD, OF NEW YORK, N. Y.

BLOCK TOY BUILDING.

1,337,171. Specification of Letters Patent. **Patented Apr. 13, 1920.**

Application filed June 18, 1919. Serial No. 304,996.

To all whom it may concern:

Be it known that I, ORIN W. WARD, a citizen of the United States, and a resident of the city of New York, borough of Manhattan, in the county of New York and State of New York, have invented a new and Improved Block Toy Building, of which the following is a full, clear, and exact description.

This invention relates to toy building constructions and has particular reference to devices of this kind being calculated to afford a vast amount of information of a practical and useful nature for boys and girls, as well as affording endless amusement in the erection of building structures.

Among the objects of the invention is to provide a base and an assemblage of interchangeable blocks having certain uniformity of character for coöperation with one another and said base so that a child may erect buildings of different designs or sizes according to his mental conception of such structures, the base being provided with groups of holes bored therein according to a definite predetermined plan for the erection of different sized buildings, using a portion of such holes at each building operation.

Another object of the invention is to provide an assembly of building units having means for interlocking different elements together in a continuous operation and with a plurality of door and window units so constructed as to coöperate with the wall units, there being provided also special means for securing the wall structures in a permanent or stable form.

With the foregoing and other objects in view the invention consists in the arrangement and combination of parts hereinafter described and claimed, and while the invention is not restricted to the exact details of construction disclosed or suggested herein, still for the purpose of illustrating a practical embodiment thereof reference is had to the accompanying drawings, in which like reference characters designate the same parts in the several views, and in which—

Figure 1 is a perspective view of a toy house built in accordance with this improvement.

Fig. 2 is a similar view but indicating portions of the elements removed or broken away and calculated to indicate the manner of assemblage and support.

Fig. 3 is a bottom plan view of one of the corner pieces such as are shown in Fig. 2.

Fig. 4 is a detail perspective view of one of the cornice elements.

Fig. 5 is a detail perspective view of a standard wall element.

Fig. 6 is a side elevation of the representation of a chimney.

Fig. 7 is a bottom plan view of the same.

Fig. 8 is a fragmentary corner portion indicating a corner post and the relation of adjacent wall elements thereto; and

Fig. 9 is a modification of a corner construction.

Referring now more specifically to the drawings, I show at 10 the base above referred to the same being provided with groups of holes 11, the holes of each group being arranged at the corner or quarter portions of the base and disposed in two straight rows at right angles to each other, one row being longitudinal of the base and the other being transverse to the base. Furthermore the longitudinal rows of holes on the same side of the base are in alinement, and the transversely arranged holes also are in alinement adjacent to the two ends of the base. Each member therefore of any group has a counterpart in each of the other groups with respect to its location or position upon the base. The purpose of these holes arranged and shown as described is to indicate to the child about to build a toy building the points necessary to locate the corners of the building in accordance with a predetermined conception of the size and form of the building and in accordance with the standard lengths of building units or elements.

The numeral 12 indicates corner posts each of which is essentially square in cross section, but provided with flanges 13 projecting at right angles from the outermost corner 14. The lower end of the post is fitted with a pin or dowel 15 adapted to be projected downward into any one of the holes 11. The upper end of each post is preferably provided with a right angled mold or cornice 16 of any suitable design.

From what has been stated above the corner posts are adapted to be set in any of the holes 11 according to the size or design of the building that the child conceives, but in any position at which the posts may be placed with respect to the symmetry of the base the distance between the corner posts will always be in accordance with the standard dimensions of the sill and wall elements.

17 indicates the base or sill members which may be either in standard short lengths so that several of them may be used to reach from one corner post to another along the same side of the building, or they may be made in various full lengths so that a single member 17 will extend all the way from one corner to the next. In either event the ends of the members 17 reach beyond or within the flanges 13 whereby outward movement of either end of the members is prevented. Furthermore one of the edges, preferably the upper edge, is provided with a tongue 18 adapted to be received in corresponding grooves 19 of the building units 20 supported upon the upper end of the sill members 17 and held from lateral displacement by means of the tongues and grooves. As above intimated the building blocks or units 20 are manufactured in uniform lengths or half lengths 20′ so that a definite number of them will exactly fill the space between adjacent corner posts providing for the breaking of joints as indicated in the drawings, thereby simulating the appearance of a standard building made of real building blocks. The walls are adapted thus to be built up to the tops of the posts.

At any desired places in the walls may be introduced window frames 21 or door frames 22. The window frames are manufactured so as to be of a predetermined height and width in accordance with the height and length of the blocks 20 and 20′. Each of the frames includes upper members or lintels 23, lower members or sills 24, as well as side members or stiles, all of the four members of each frame being assembled in the factory and so are maintained as fixed units. The uppermost and side members of the frames are provided with parallel outwardly projecting flanges 25 which receive the adjacent portions of the block members 20 and 20′ and so hold the block members from lateral movement. The door 26 may be carried as a hinged portion of the frame 22. The door and window frames are likewise held from movement laterally with respect to the wall members and consequently these members may all be said to have an interlocking action.

Any suitable means may be provided to make permanent or secure the interlocking action between the corner posts and the wall members adjacent thereto. In Figs. 2 and 3 I show for this purpose a pair of plates 27 secured upon the inner faces of each post

and extending therefrom in vertical planes at right angles to each other, each plate being in position to bear directly against the inner face of one or more of the block elements 20 or 20′ whereby such elements are positively held from movement inward from normal vertical position. As shown in Fig. 2 one member 27 of each pair is mated with a corresponding member of the next adjacent post, being located at the same level as the mating member, and between such two mating members 27 I extend a tie member 28 shown as comprising two chains connected by a coil spring 29. Certain links of the chain members may be hooked over hooks 30 carried by the plates 27. The flexibility of the spring 29 provides for the putting of the tie members under sufficient tension to make a rigid binding of the parts together. This serves to prevent the outward tilting of the upper ends of the posts and without such tilting there can be no separation between any two adjacent portions of the structure.

As an alternative form of corner structure and interlocking means I show in Fig. 8 a corner comprising a post 12′ having outside flange members 13′ with the same function as ascribed to the flanges 13 above, but in place of the plates 27 I provide a structure comprising a single plate 31 of sheet metal fastened permanently as by means of nails or screws 32 to one inner face of the post. The width of this plate 31 is sufficient to provide extensions 33 in the same plane as the main portion of the plate and spaced from each other while the portions 34 of the main plate lying between the projections 33 are bent from the post or the plane of the plate 31 into another plane at right angles to the plate 31 so as to be in position to prevent the inward displacement of the wall members abutting against the plate 31. The projections 33 obviously will prevent inward displacement of the wall members lying between them and that flange projecting parallel to the members 33. In some constructions this form of post shown in Fig. 8 is quite sufficient to prevent any separation of the parts of the building and such structure is simpler and cheaper than the tie means indicated in Fig. 2. The wall elements 20 coöperate with these posts the same as in the other structure. Before proceeding further with the roof or top portions of the building I call attention at this time to the alternative corner structure in Fig. 9 of which special wall elements are provided for the corners, said elements 20ᵃ being each provided with a tongue 18′ along its upper edge similar in location and function to the tongues 18 of the standard elements 20, but the bottom of each element 20ᵃ is provided with a groove 19′ extending transversely of the block or element and hence at right angles to the tongue 18′ of the same element. Consequently the groove 19′ of any element has a direct interlocking coöperation with the tongue 18′ of the next lower element. Therefore no block or element can be displaced in any direction without disturbing the entire combination. This form of interlocking corner is very efficient for toy building purposes and carrying out the general spirit of the invention affords boundless amusement as well as instruction to the child.

Supplemental to and harmonious with the corner post cornices 16 I provide any number of cornice members or blocks 20ᶜ each fitted with a cornice portion 16′. These special elements 20ᶜ with the attached cornices are illustrated in Fig. 4 and in assembly in Figs. 1 and 2. Roof or ceiling members are shown at 35 in the form of rectangular slabs adapted to extend across the building and rest upon the ledges provided just inside of the tongues 18 of the uppermost elements. These slabs may be provided with suitably arranged holes in any one of which a pin 36 extending downward from a chimney member 37 is adapted to project.

I claim,

1. The herein described toy building comprising a plurality of interlocking wall blocks or elements, vertical corner posts each having laterally projecting flanges on its outer sides adapted to receive and hold the end portions of the adjacent wall members preventing outward displacement of such members with respect to the posts, and flexible tie means extending along each wall from one post to the next serving to prevent outward tilting of the posts and consequent separation of the wall members therefrom.

2. In a toy building of the character set forth, the combination of a plurality of spaced corner posts, a base member with which said posts have interlocking connection preventing lateral movement of the lower end of the posts and separable wall elements adapted to be built up between the corner posts, of tie means extending from each post to the next along the walls, said tie means including mating pairs of plates carried by adjacent corner posts, and a flexible member extending from one plate to the other and adjustably secured thereto.

3. In a toy building, the combination with a base having a plurality of holes formed therein, of a plurality of corner posts having dowels projecting outward therefrom into certain of said holes, wall elements built up between corner posts and having interlocking engagement therewith, and flexible and resilient tie means extending from one post to the next serving to prevent outward tilting of the upper ends of the post and releasing of the uppermost wall elements.

4. In a toy building construction, a corner post having a plurality of plates projecting from the inner sides of the posts in vertical planes at right angles to each other, and flexible tying means connecting said plates so as to prevent the outward tilting of said posts.

ORIN WARREN WARD.

- 3 -

Dipl.-Ing. Manfred Vielmo
Stuttgart-Sonnenberg

20. März 1967

13. Puppenhaus nach einem der Ansprüche 7 bis 12, dadurch gekennzeichnet, daß die Höhe wenigstens einer Tragwand (2) doppelt
so groß ist wie die Länge der Stockwerksplatten (1) und daß
die übrigen Tragwände (2) um die Höhe der Trennwände (11)
niedriger sind als die höchste Tragwand (2).

Bek. gem. 2 1. Sep. 1967

77f. 3/52. 1 969 017. Dipl.-Ing. Man-
fred Vielmo, Stuttgart-Sonnenberg. |
Mehrstöckiges Puppenhaus.   30. 3. 67.
V 20 182. (T. 7; Z. 2)

Dipl.-Ing. Manfred Vielmo
7 Stuttgart-Sonnenberg
Kießstraße 15

P.A. 173 408 * 30. 3. 67   23. März 1967

An das
Deutsche Patentamt

8000 München 2
Zweibrückenstr. 12

Neue Gebrauchsmusteranmeldung

Hiermit melde ich,

Manfred Vielmo, Dipl.-Ing., Stuttgart-Sonnenberg, Kießstr. 15,

den in den Anlagen beschriebenen Gegenstand als Gebrauchsmuster an
und beantrage seine Eintragung in die Rolle.

Die Bezeichnung lautet:

Mehrstöckiges Puppenhaus.

Die Anmeldegebühr in Höhe von DM 30,-- wird auf das Postscheck-
konto München Nr. 791 91 des Deutschen Patentamtes überwiesen werden, sobald mir das Aktenzeichen bekannt ist.

Anlage
1. Zwei weitere Ausfertigungen dieses Antrags,
2. drei gleichlautende Beschreibungen mit je 13 Ansprüchen,
3. drei Satz Aktenzeichnungen mit je 2 Blatt
4. eine vorbereitete Empfangsbescheinigung.

Aussetzungsantrag:
Ich beantrage ferner, die Eintragung des Gebrauchsmusters um
sechs Monate auszusetzen.

Manfred Vielmo
(Manfred Vielmo)

Nr. 1 969 017 * eingetr.
2. 1. 9 67

P.A. 173408 * 30. 3. 67

Dipl.-Ing. Manfred Vielmo   23. März 1967
7 Stuttgart-Sonnenberg
Kießstraße 15

Anlage zur
Patent- und Gebrauchsmusteranmeldung

Mehrstöckiges Puppenhaus

Die Erfindung bezieht sich auf ein auf- und abbaubares, mehrstöckiges Puppenhaus mit Tragwänden und Stockwerksplatten, die aneinander durch Steckverbindungen befestigt sind.

Derartige Puppenhäuser sind bereits bekannt. Sie sind gewöhnlich mit einem Boden versehen, der als Grundplatte für das Puppenhaus dient. Das hat den Nachteil, daß diese Häuser, da sie bei kleinsten Bodenunebenheiten keinen festen Stand haben, wackeln; außerdem können Spielzeugautos nicht in das Puppenhaus eingefahren werden, ohne daß eine Rampe o. dgl. vorhanden ist. Da ein derartiges Puppenhaus nur nach einer Seite hin offen ist, kann es höchstens von zwei Kindern bespielt werden, wobei diese beiden Kinder sich aber bereits gegenseitig behindern. Schließlich ist zum Zusammen- oder Auseinanderbauen des Puppenhauses oft Handwerkszeug nötig, mit dem viele Kinder nicht richtig umzugehen verstehen.

Aufgabe der vorliegenden Erfindung ist es, diese Nachteile zu vermeiden und ein Puppenhaus zu schaffen, das besonders standfest ist, das von mehreren Seiten bespielbar ist und das ohne Werkzeuge auf- und abgebaut werden kann.

Diese Aufgabe wird gemäß der Erfindung gelöst durch mehrere durchgehende Tragwände, die bodenfrei aufgestellt sind und in Stockwerkshöhe waagrechte Schlitze aufweisen, in welche die eine gesamte Stockwerksfläche bildenden Platten eingreifen.

Eine besonders vorteilhafte Bauart des erfindungsgemäßen Puppenhauses wird gemäß einem weiteren Merkmal dadurch erreicht, daß

- 2 -

Dipl.-Ing. Manfred Vielmo   23. März 1967
Stuttgart-Sonnenberg

die Stockwerksplatten ebenfalls mit Schlitzen versehen sind, in die der durchgehende Teil der Tragwände eingreift. Dadurch ist ein sehr leichter, jedem Kind zumutbarer Auf- und Abbau des Puppenhauses möglich.

Des weiteren ist es gemäß einem anderen Merkmal des Puppenhauses vorteilhaft, daß für jedes Stockwerk auswechselbare, nicht tragende Trennwände vorgesehen sind, die in Nuten der Stockwerksplatten geführt sind. Dadurch ist es möglich, die Räume im Puppenhaus nach Belieben immer wieder zu verändern, was dem gestaltenden Spiel besonders dienlich ist. Geschlossene Wände können z.B. auch gegen Wände mit Fenster oder Türen ausgetauscht werden.

Schließlich ist es von Vorteil, daß gemäß einem weiteren Merkmal der Erfindung die Höhe wenigstens einer Tragwand doppelt so groß ist wie die Länge der Stockwerksplatten und daß die übrigen Tragwände um die Höhe der Trennwände niedriger sind als die höchste Tragwand. Dadurch ist es möglich, das Puppenhaus zu einem handlichen Paket zusammenzulegen, das leicht aufbewahrt und überallhin, z.B. in den Urlaub, mitgenommen werden kann.

Ein Ausführungsbeispiel der Erfindung ist in der Zeichnung dargestellt; es zeigen:

Fig. 1 ein aufgestelltes Puppenhaus in raumbildlicher
        Darstellung,
Fig. 2 eine Seitenansicht und
Fig. 3 einen Schnitt durch ein Stockwerk des Hauses.

Das freistehende Puppenhaus in Form eines Hochhauses hat beispielsweise drei Stockwerke, deren Decken und Böden von vier quadratischen Platten 1 gleicher Grundfläche gebildet werden. In Stockwerkshöhe werden die Platten 1 von vier auf dem Boden aufgestellten, durchgehenden Tragwänden 2 in Form einer Steck- oder Kammverbindung gehalten. Dazu haben die Tragwände 2 in Stockwerkshöhe Schlitze 3 (vgl. Fig. 3), deren Länge etwa die Hälfte der

- 3 -

- 3 -

Dipl.-Ing. Manfred Vielmo   23. März 1967
Stuttgart-Sonnenberg

Breite der jeweiligen Tragwand beträgt und deren Breite gleich
der Dicke der Platten 1 ist. Entsprechend sind in die Platten 1
rechtwinklig vom inneren Begrenzungsrand ausgehende Schlitze 4
eingeschnitten, deren Länge etwa gleich der der Schlitze der zugeordneten Tragwände 2 und deren Breite gleich der Dicke derselben ist. Die Passung der Schlitze 3 und 4 ist so gewählt, daß
die Platten 1 und Tragwände 2 leicht ineinanderschiebbar sind
und daß die Steckverbindung dennoch rüttelfest ist.

Die Tragwände 2 haben unterschiedliche Breiten, so daß zwischen
ihnen als Türöffnungen dienende Durchgänge freibleiben (vgl. Fig. 3).
Während zwei der Tragwände 2 in derselben Ebene liegen, sind die
anderen beiden gegeneinander versetzt. Es entstehen dadurch Räume
unterschiedlicher Größe.

Um die Festigkeit der Tragwände 2 zu erhöhen und um eine hohe
Standfestigkeit des Puppenhauses zu gewährleisten, überragen die
Tragwände 2 die Platten 1 seitlich. Bei einer der Tragwände 2 ist
der überragende Teil 5 so breit gehalten, daß ein Aufzug daran geführt werden kann. Die Aufzugkabine 7 ist an einem endlosen Seil 9
befestigt, das auf Rollen 6 läuft, die in Aussparungen 16 im oberen
und unteren Ende der betreffenden Tragwand 2 drehbar auf Bolzen 8
gelagert sind.

Auf ihrer Ober- und Unterseite haben die Platten 1 von einer Seite her bis zu dem jeweiligen Schlitz 4 in geringem Abstand parallel
zum Begrenzungsrand verlaufende Nuten 10, in die Trennwände 11 eingeschoben sind oder eingeschoben werden können (vgl. Fig. 3). Diese
Trennwände 11 können als Fenster dienende Durchbrüche 12 haben.
Weitere Nuten 13 sind in einem größeren Abstand von einem Begrenzungsrand der Platten 1 angeordnet, in die Trennwände 14 mit einer
als Tür dienenden Aussparung 15 eingeschoben sind oder eingeschoben
werden können (vgl. Fig. 1).

Damit das Puppenhaus in auseinandergebautem Zustand nur geringen
Platzbedarf benötigt, ist wenigstens eine Tragwand 2 doppelt so

- 4 -

Dipl.-Ing. Manfred Vielmo   23. März 1967
Stuttgart-Sonnenberg

hoch wie eine Platte 1 lang. Die anderen Tragwände sind um die
Höhe oder Breite einer Trennwand niedriger. Auf diese Weise können
die Einzelteile des Puppenhauses so gestapelt werden, daß zuunterst
zwei Tragwände 2, darauf in zwei Schichten jeweils nebeneinander
zwei Platten 1 und schließlich die restlichen kürzeren Tragwände 2
neben den Trennwänden 11 und 14 liegen.

Das Puppenhaus wird wie folgt aufgebaut:
In die Schlitze 3 einer der Länge nach auf den Boden gelegten, mit
den Schlitzen 3 nach oben ragenden Tragwand 2 werden nacheinander
die Platten 1 mit ihren zugeordneten Schlitzen 4 eingesteckt. Darauf
wird von oben her die gegenseitige Tragwand 2 mit ihren Schlitzen 3
in die der Platten 1 eingeschoben. Auf gleiche Weise werden dann
die zwei anderen Tragwände 2 von der Seite her mit den Platten 1
verbunden. Das Ganze wird umgekippt und aufgestellt. Danach können
die Trennwände 11 und 14 nach Belieben in die Nuten 10 und 13 eingeschoben werden.

P.A. 173408 * 30.3.67

Dipl.-Ing. Manfred Vicino
Stuttgart-Sonnenborg

29. März 1967

Schutz-Ansprüche

1. Aus- und abbaubares, mehrstöckiges Puppenhaus mit Tragwänden und Stockwerksplatten, die aneinander durch Steckverbindungen befestigt sind, gekennzeichnet durch mehrere durchgehende Tragwände (2), die bodenfrei aufgestellt sind und in Stockwerkshöhe waagerechte Schlitze (3) aufweisen, in welche die eine gesamte Stockwerksfläche bildenden Platten (1) eingreifen.

2. Puppenhaus nach Anspruch 1, dadurch gekennzeichnet, daß die Stockwerksplatten (1) ebenfalls mit Schlitzen (4) versehen sind, in die der durchgehende Teil der Tragwände (2) eingreift.

3. Puppenhaus nach Anspruch 1 oder 2, dadurch gekennzeichnet, daß die Tragwände (2) zwischen einander Durchgänge freilassen.

4. Puppenhaus nach einen der Ansprüche 1 bis 3, dadurch gekennzeichnet, daß die Tragwände (2) rechtwinklig zueinander angeordnet sind.

5. Puppenhaus nach einem der Ansprüche 1 bis 4, dadurch gekennzeichnet, daß die Tragwände (2) in mehr als zwei Ebenen angeordnet sind.

6. Puppenhaus nach einem der Ansprüche 1 bis 5, dadurch gekennzeichnet, daß die Tragwände (2) die Stockwerksplatten (1) seitlich überragen.

Dipl.-Ing. Manfred Vicino
Stuttgart-Sonnenborg

29. März 1967

7. Puppenhaus nach (einem der) Ansprüche 1 bis 6, dadurch gekennzeichnet, daß für jedes Stockwerk auswechselbare, nichttragende Trennwände (11, 14) vorgesehen sind, die in Nuten (10, 13) der Stockwerksplatten (1) geführt sind.

8. Puppenhaus nach Anspruch 7, dadurch gekennzeichnet, daß die Trennwände (11) volle Platten sind.

9. Puppenhaus nach Anspruch 7, dadurch gekennzeichnet, daß die Trennwände (11, 14) Durchbrüche, insbesondere Fenster- und Türöffnungen, aufweisen.

10. Puppenhaus nach einem der Ansprüche 1 bis 9, dadurch gekennzeichnet, daß die Stockwerke in Griffhöhe von Kindern bespielbar sind.

11. Puppenhaus nach einem der Ansprüche 1 bis 10, dadurch gekennzeichnet, daß eine Tragwand (2) den Rand der Stockwerksplatten (1) wesentlich überragt und daß an den überragenden Teil ein Aufzug (6, 7, 8, 9) vorgesehen ist.

12. Puppenhaus nach einem der Ansprüche 7 bis 11, dadurch gekennzeichnet, daß die Stockwerksplatten (1) gleiche Grundflächen und gleiche Anordnung der Schlitze (4) und Nuten (10) aufweisen.

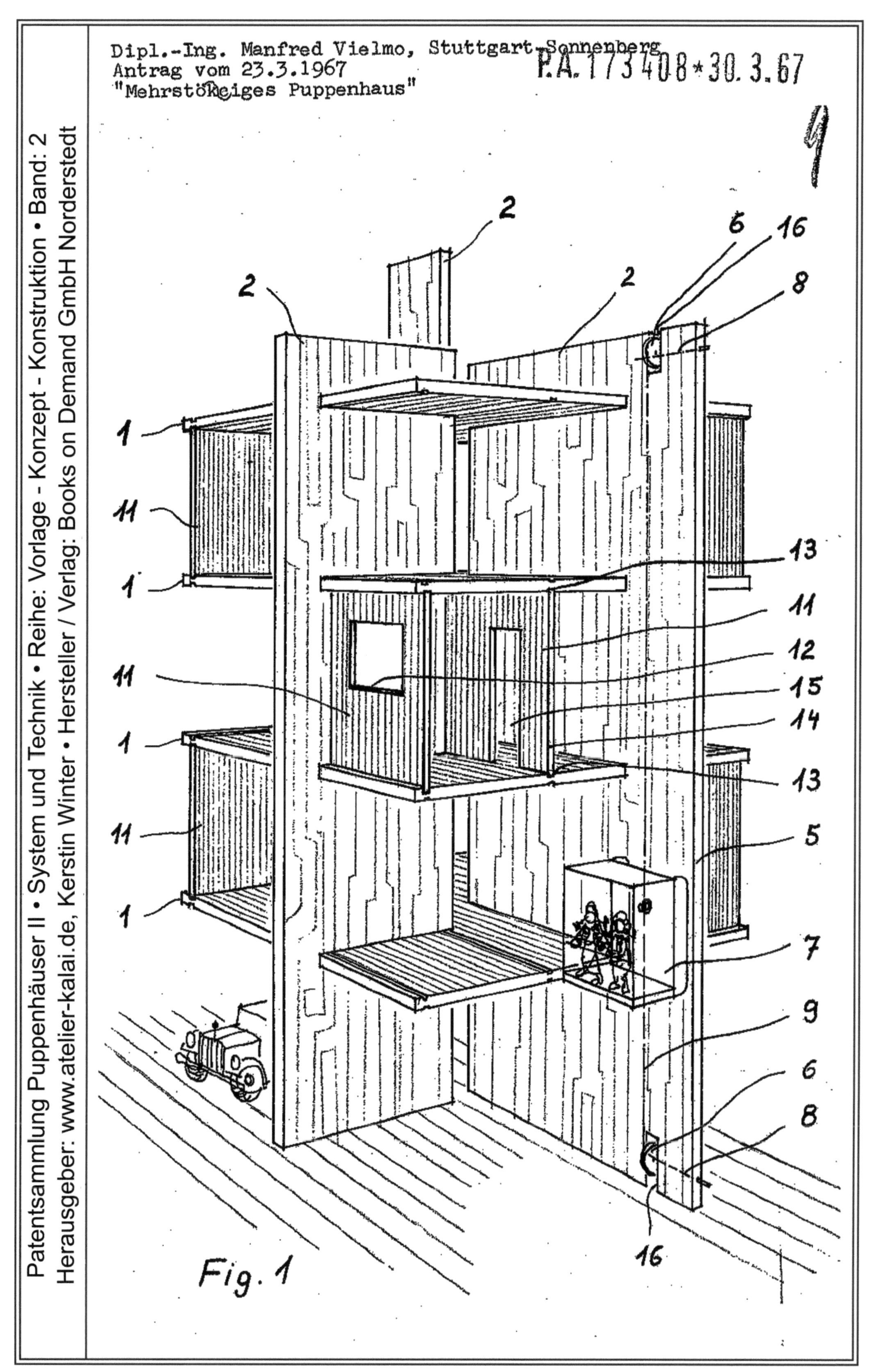

Fig. 1

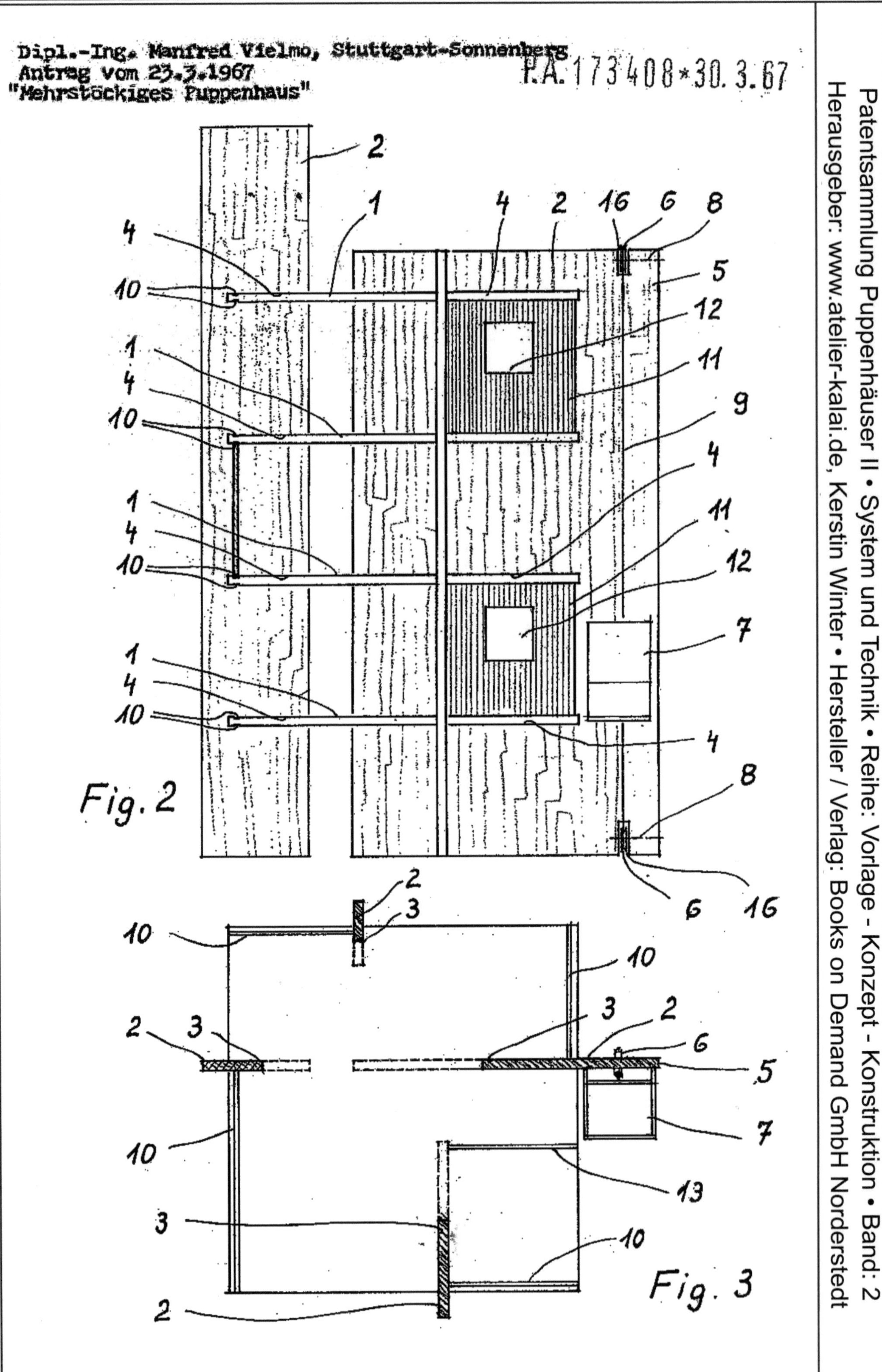

Dipl.-Ing. Manfred Vielmo, Stuttgart-Sonnenberg
Antrag vom 23.3.1967
"Mehrstöckiges Puppenhaus"
P.A. 173 408 * 30. 3. 67
Fig. 2
Fig. 3

H. DE L. RAPSON.
COLLAPSIBLE TOY HOUSE.
APPLICATION FILED OCT. 14, 1920.

1,386,423.

Patented Aug. 2, 1921.

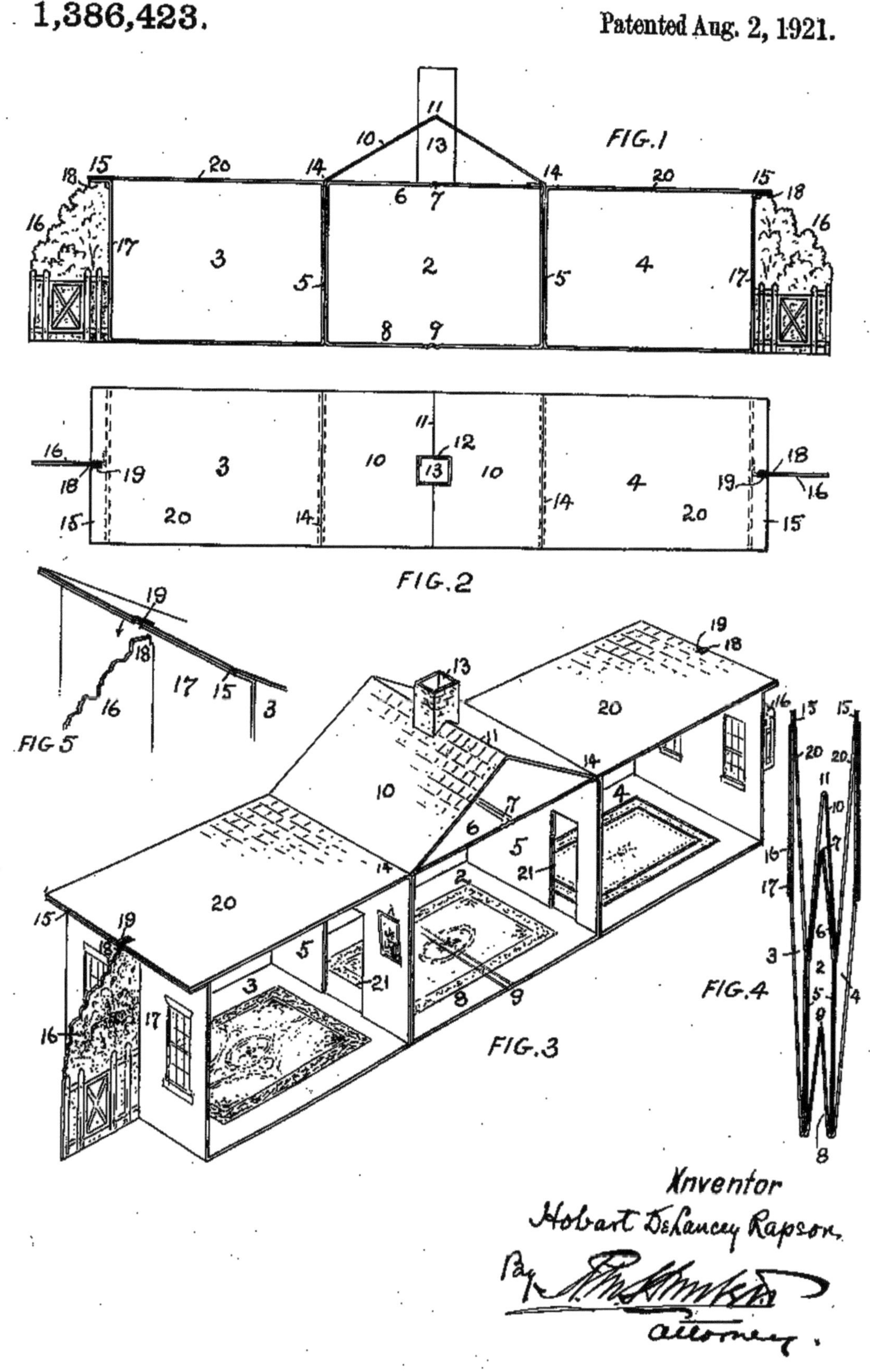

UNITED STATES PATENT OFFICE.

HOBART DE LANCEY RAPSON, OF PHILADELPHIA, PENNSYLVANIA.

COLLAPSIBLE TOY HOUSE.

1,386,423.   Specification of Letters Patent.   **Patented Aug. 2, 1921.**

Application filed October 14, 1920. Serial No. 416,976.

To all whom it may concern:

Be it known that I, HOBART DE LANCEY RAPSON, a citizen of the United States, and resident of Philadelphia, county of Philadelphia, State of Pennsylvania, have invented an Improvement in Collapsible Toy Houses, of which the following is a specification.

The object of my invention is to provide a construction of collapsible doll's house which shall embody simplicity and cheapness and which, moreover, permits quick erection with accuracy of adjustment of the parts relatively to each other.

My object is further to provide a construction of a cardboard doll's house which will provide a number of rooms which are open on opposite sides so that children thereat can coöperatively play.

My object is further to provide a multiroom house so constructed that all of the rooms may be brought to the same plane but which may be collapsed to provide a closely compacted cardboard structure which occupies relatively little space and which may be packed and shipped without the least danger of injury.

In the preferred construction of my improved toy house, the various surfaces are printed in colors to represent the roofing, the outside windows and masonry work, the rugs and general wall and ceiling decorations on the interior, all of which gives to the entire structure not only an ornamental appearance suitable for the purpose for which the article is employed, but also imparts to the structure an appearance of being already half furnished before the toy furniture, etc., is added.

For the purpose of illustrating my invention I have shown in the accompanying drawings the embodiment thereof which is at present preferred by me, since the same is in form to give satisfactory and reliable results, but it is to be understood that the several instrumentalities of which my invention consists can be variously arranged and organized and that my invention is not limited to the precise arrangement and organization of the instrumentalities herein shown and described.

Referring to the drawings: Figure 1 is a side elevation of my improved collapsible toy house in its erected condition; Fig. 2 is a plan view of the same; Fig. 3 is a perspective view of the same; Fig. 4 is a side view illustrating the device in the process of being completely collapsed; and Fig. 5 is a perspective view illustrating a detail.

In general, the structure is composed of cardboard or other material bent in definite shapes and glued or stapled together so that the entire structure is practically integral, but with capacity of being collapsed by flexibility at certain of the bends or joints. 2 is a rectangular cardboard structure open at opposite ends, but providing a ceiling portion 6 and a floor portion 8. In addition to its rectangular form, it is creased transversely at 7 in the ceiling portion 6 and at 9 in the floor portion 8, said creases permitting the said parts 6 and 8 to fold upwardly in the middle when bringing the side portions 5 together, as will be understood by reference to Fig. 4. When this structure 2 is extended, it is rectangular, as shown in Fig. 1. In addition to the rectangular part 2, there are similar rectangular parts 3 and 4 which are formed of bent cardboard strips with the free edges united, as at 15, said united parts forming a two-ply rigid overhanging portion which might be considered as the eaves of a roof. These rectangular parts 3 and 4 are respectively placed at opposite ends of the part 2 and glued to them, as indicated adjacent to the walls 5.

In addition to these three rectangular parts 2, 3 and 4, there is an additional inverted V shaped part 10 which constitutes a peaked roof having the ridge 11. The lower ends of the roof structure 10 are glued between the abutting sides of the parts 2, 3 and 4, as indicated at 14. It will be understood that when the part 2 is folded by bringing the side walls 5, 5, toward each other and with the hinged parts 7 and 9 move upward, the ridge 11 of the roof 10 also acts as a hinge as the roof folds, as is clearly indicated in Fig. 4. To heighten the effect of the roof structure, I prefer to provide the roof part 10 with a rectangular aperture 12 through which is inserted a rectangular chimney part 13 which not only gives the appearance of a chimney but also acts as a strut to hold the ceiling 6 horizontal and give somewhat more rigidity. This chimney may snugly fit the aperture 12, so as to provide a frictional contact therewith. The side walls 5, being double ply, are quite rigid and give considerable strength to the structure as a whole and these walls may be provided with doorways

Patentsammlung Puppenhäuser II • System und Technik • Reihe: Vorlage - Konzept - Konstruktion • Band: 2 Herausgeber: www.atelier-kalai.de, Kerstin Winter • Hersteller / Verlag: Books on Demand GmbH Norderstedt

21 (Fig. 3), if so desired, to still further heighten the appearance of the house structure and also permit the passing of the dolls from one room to another.

The roof part 10, as well as the top portions 20 of the room compartments 3 and 4, are printed to represent shingles or roofing structure and while the parts 2, 3 and 4 constitute rooms, the part above the ceiling 6 and inclosed by the peaked roof 10 constitutes an attic, which still further provides suggestions for the children's play. The construction of the parts 3 and 4 is such that all of the four corners constitute hinges so that at the same time that the central part 2 collapses sidewise, somewhat on a bellows principle, the side portions 3 and 4 are collapsed by swinging upward, as is clearly indicated in Fig. 4.

While the foregoing describes the general structure of the house proper, additional stability and artistic appearance is insured by the provision of the end pieces 16 which are hinged to the vertical outer wall 17 of the room portions 3 and 4. These parts 16 are preferably printed to represent fence and gate structures with foliage and their upper ends 18 project sufficiently to engage notches 19 in the overhanging eave portions 15 when the collapsed room portions 3 and 4 are lowered in bringing the structure to the erected position, all of which will be understood by reference to Figs. 3 and 5. When the part 18 is engaged in the notch 19, the part 16 will be at right angles to the wall 17 and will assist in giving rigidity to the ends of the erected structure as a whole, and this is desirable, since the walls 17 are of single ply and consequently liable to become creased and broken. The parts 16 act as braces for these wall portions 17 and also aid by their weight in holding the said parts down to a horizontal position when the device is erected. It will be further understood that when the structure is collapsed, these parts 16 are folded flat against the face of the wall portions 17, so that the whole collapsed structure takes very little space and may be easily transferred through the mails or packed for shipment without any danger of injury.

From the general construction shown and described, it will be apparent that as there are no sides to the house structure when erected, the whole device is so constituted that all of the main parts swing in the same plane and hence may be quickly collapsed and as quickly extended into erected form, and the said operations are so simple that a child may readily perform the necessary operations. It will also be seen that by having no sides to the building structure, all of the interior is readily exposed at either side for the insertion of the hands of the children for manipulating the dolls and furniture, which not only simplifies the use of the article, but also provides against injury thereto which would result by difficult entrance of the hands. Furthermore, the manner in which the parts 3 and 4 are hinged to the central part 2 operates to hold the central part in an extended position, said parts 3 and 4 acting as levers and weighted portions to pull the ceiling portion 6 horizontal and in extended form.

I have shown in the construction as an embodiment of my invention the features which I have found most preferable, but I do not confine myself to the particular manner of jointing or hinging the parts which provide a plurality of rooms in an erected toy house formed of cardboard, as the same general results and advantages may be secured by modifications of the general invention involving the collapsible improvements.

Having now described my invention, what I claim as new and desire to secure by Letters Patent, is:—

1. A one story collapsible toy house, consisting of a plurality of rectangular box-like frames open at each end and permanently connected together side by side on parallel hinge parts to provide a plurality of rooms in the same horizontal alinement, said box-like frames arranged to fold partly in opposite directions toward each other into a flat condition with all of the parts constituting the top, bottom and sides thereof collapsible so as to lie close together in parallel relation.

2. A collapsible toy house consisting of three rectangular frames open at each end and connected together to provide a plurality of rooms in the same horizontal plane, said frames arranged to fold into a flat condition with all of the parts constituting the top, bottom and sides thereof collapsible so as to lie close together in parallel relation, and in which the middle frame has its top and bottom parts formed with hinge portions in their middle, whereby these parts each fold upon themselves when the side parts are brought together.

3. The invention according to claim 2, wherein the top of the middle frame is provided with a raised roof portion normally formed in inverted V shape and which is also adapted to be collapsed with said roof portion folded upon itself as in the case of the top and bottom parts of said middle frame.

4. The invention according to claim 3, wherein the roof portion of the middle frame is provided with an aperture at the ridge part thereof and combined with a vertical detachable chimney portion arranged to extend through the aperture and at its inner end resting upon the top part of the middle frame.

5. The invention according to claim 2, wherein the outer rectangular frames are formed as collapsible trapezoids and swing toward each other and upon the middle collapsible frame.

6. A collapsible toy house consisting of three rectangular frames open at each end and connected together to provide a plurality of rooms in the same horizontal plane, said frames arranged to fold into a flat condition with all of the parts constituting the top, bottom and sides thereof collapsible so as to lie close together in parallel relation, and in which the outer sides of the two end frames are provided with brace parts hinged thereto and having interlocking portions engaging the upper parts of said frames when opened in erected form.

7. A rectangular collapsible toy house structure, comprising a strip of cardboard bent into rectangular form with hinge portions at each corner and open at each end, the top and bottom parts also creased along their middle portions so that they may each be folded upon itself and the sides brought together, said structure also having a roof formed of cardboard of inverted V form having its edges united to the sides of the structure and adapted to fold upon itself when the structure is collapsed.

8. A collapsible toy house, consisting of a plurality of rectangular box-like frames permanently open at opposite ends, said frames connected together side by side to provide a plurality of integrally united rooms in the same horizontal alinement, each of said frames arranged to fold into a flat condition and in which all of the parts constituting the top, bottom and sides of the plurality of box-like frames are collapsible so as to lie close together in parallel relation.

In testimony of which invention, I hereunto set my hand.

HOBART DE LANCEY RAPSON.

Patentsammlung Puppenhäuser II • System und Technik • Reihe: Vorlage - Konzept - Konstruktion • Band: 2
Herausgeber: www.atelier-kalai.de, Kerstin Winter • Hersteller / Verlag: Books on Demand GmbH Norderstedt

Jan. 17, 1961   N. I. PAULSON   2,968,118

TOY BUILDING BLOCKS

Filed June 18, 1956   3 Sheets–Sheet 1

FIG. 1

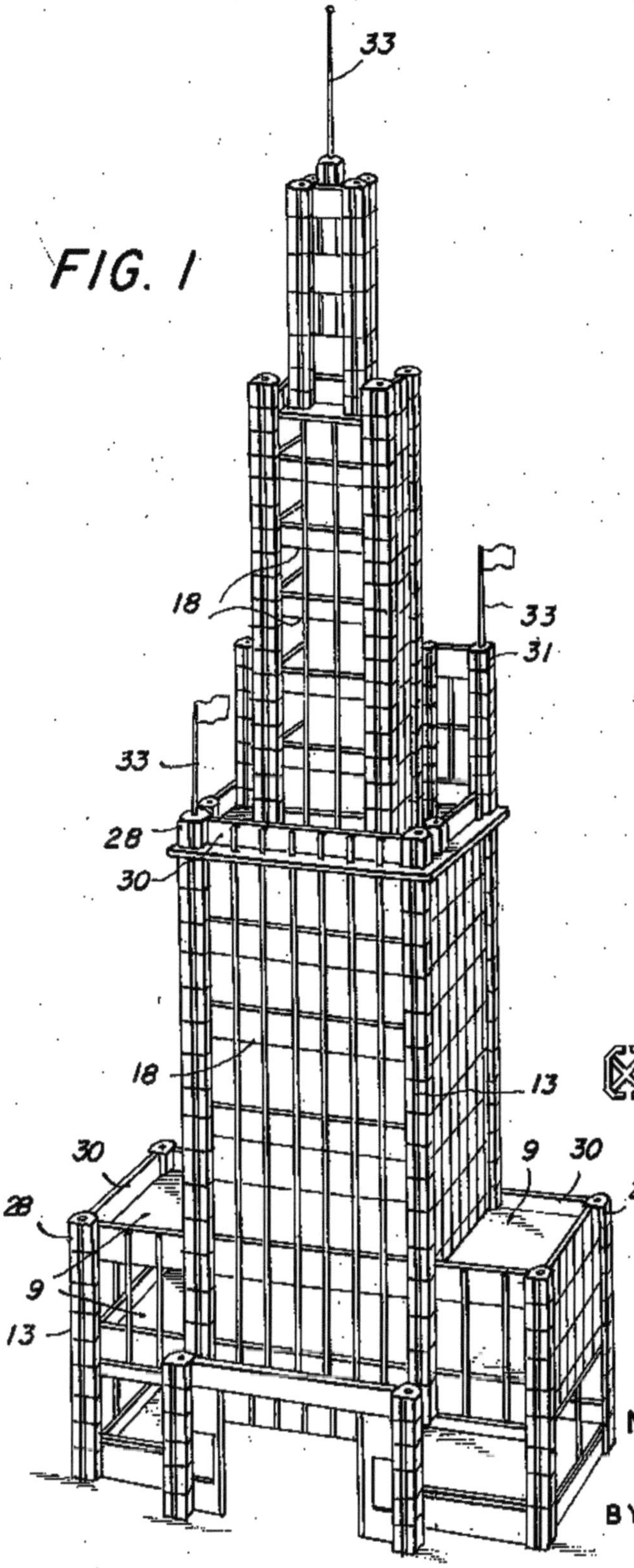

FIG. 12

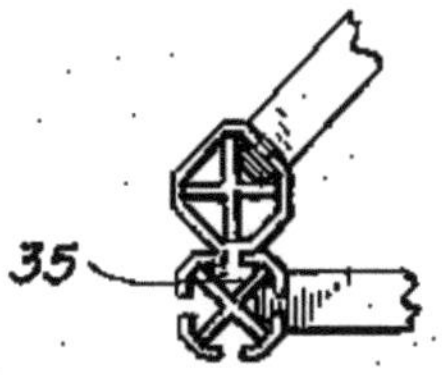

FIG. 13

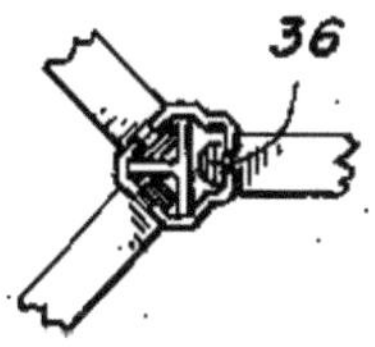

FIG. 14

INVENTOR

NILS I. PAULSON

BY

ATTORNEY

Patentsammlung Puppenhäuser II • System und Technik • Reihe: Vorlage - Konzept - Konstruktion • Band: 2
Herausgeber: www.atelier-kalai.de, Kerstin Winter • Hersteller / Verlag: Books on Demand GmbH Norderstedt

Jan. 17, 1961

N. I. PAULSON

2,968,118

TOY BUILDING BLOCKS

Filed June 18, 1956

3 Sheets-Sheet 2

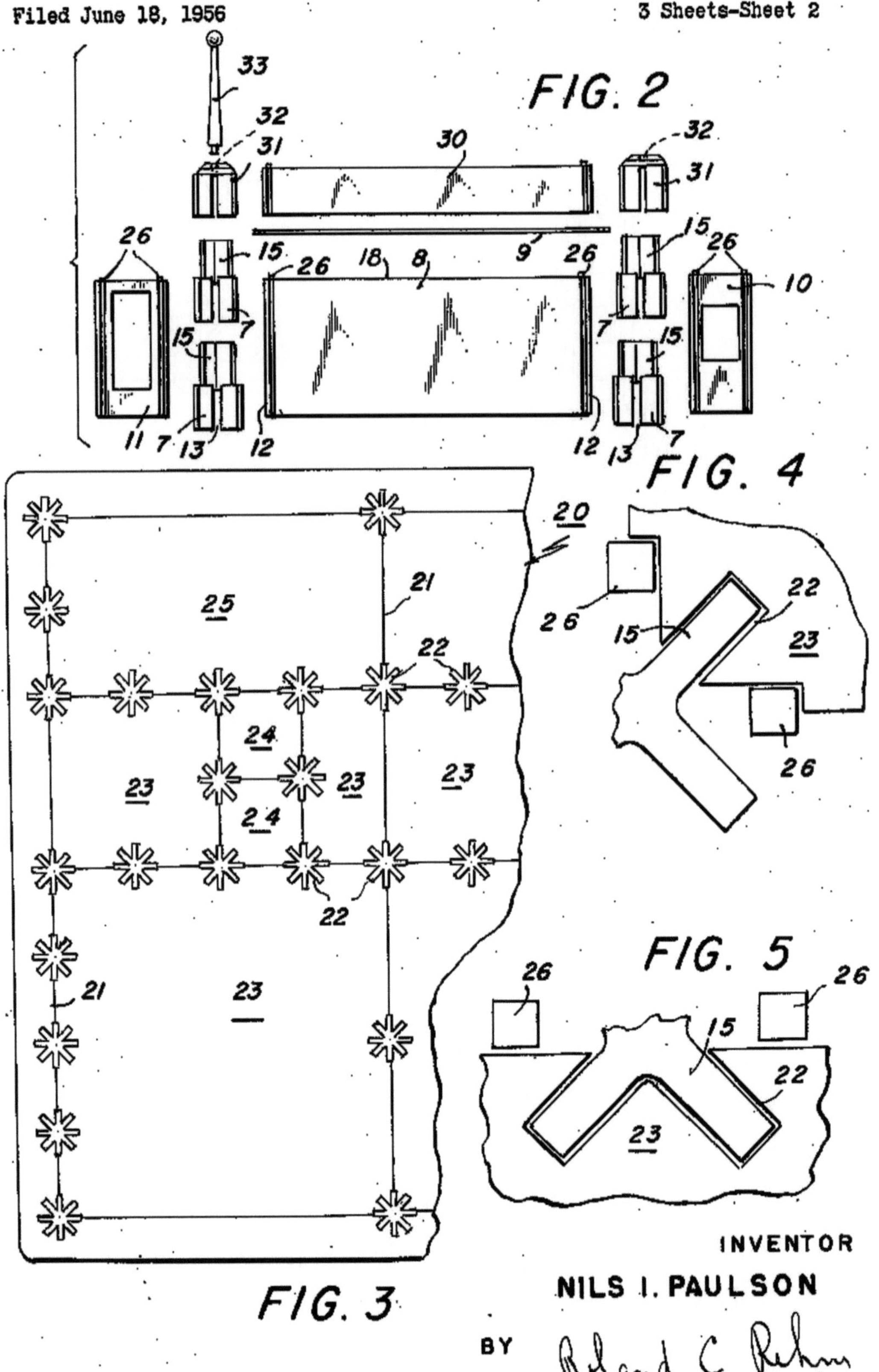

Jan. 17, 1961
N. I. PAULSON
2,968,118
TOY BUILDING BLOCKS
Filed June 18, 1956
3 Sheets-Sheet 3
FIG. 6
FIG. 11
FIG. 7
FIG. 8
FIG. 9
FIG. 10
INVENTOR
NILS I. PAULSON
BY
ATTORNEY

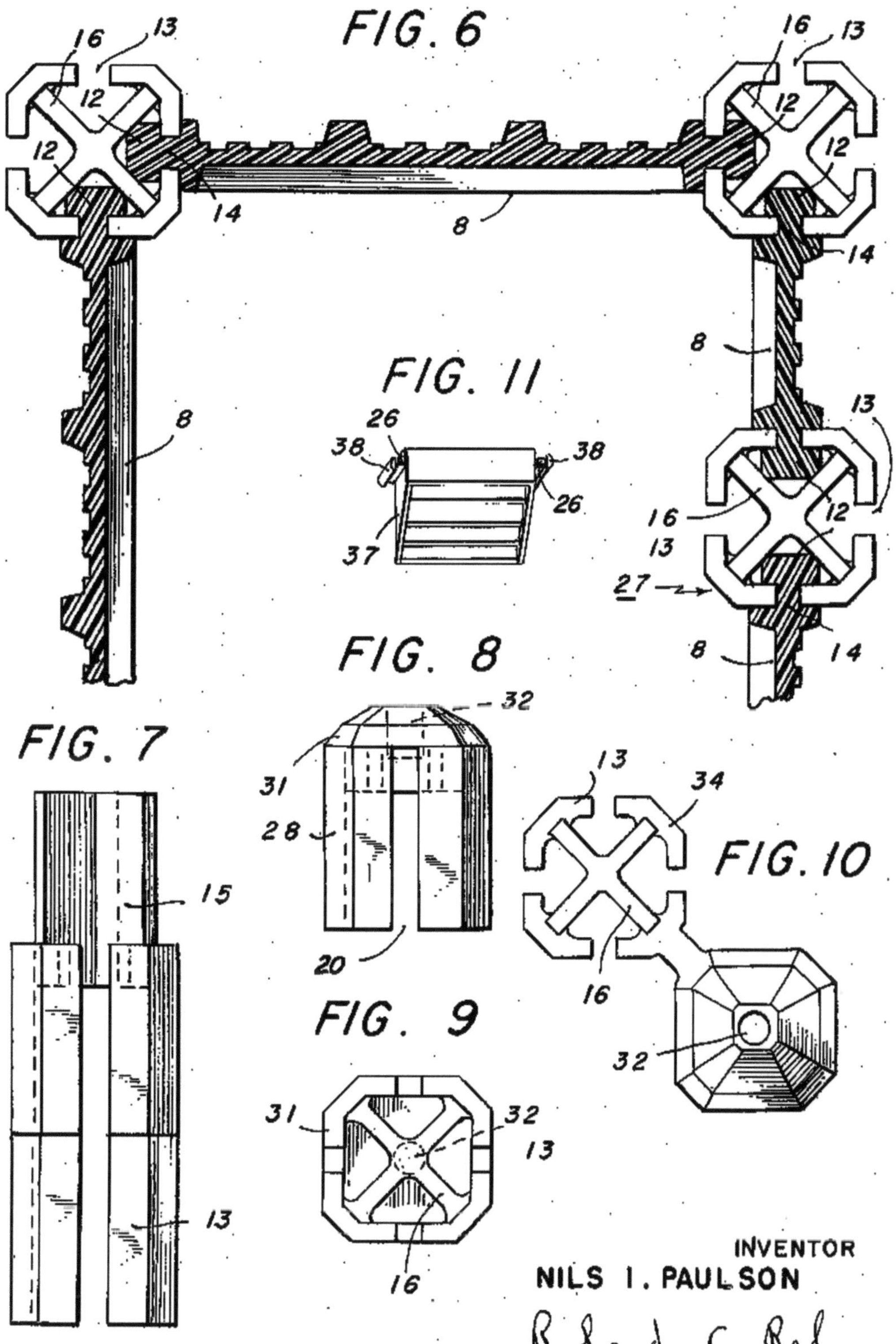

1

2

2,968,118

TOY BUILDING BLOCKS

Nils I. Paulson, Chicago, Ill., assignor to Halsam Products Company, Chicago, Ill., a corporation of Illinois

Filed June 18, 1956, Ser. No. 591,960

6 Claims. (Cl. 46—19)

This invention relates to toy building blocks.

The several building elements are designed to make it possible to simulate, if desired, conventional and multi-storied building structures.

The nature of the invention and further details thereof will readily appear from the following description of illustrative building elements embodying the invention and shown in the accompanying drawings.

In said drawings:

Fig. 1 is a perspective view of an illustrative multistoried building made from the illustrative building blocks or structural elements or units;

Fig. 2 is an "exploded" elevation of illustrative structural elements detached from each other but having the relative arrangement which they occupy when assembled;

Fig. 3 is a plan view of a blank used for floors and formed with weakened or "break" lines to facilitate subdivision of the blank into floor sections of the desired size and shape;

Fig. 4 is an enlarged plan view of the corner of a floor section at a column and showing the spacing pads which vertically spaced superposed units by the thickness of the floor;

Fig. 5 is a similar detail of a floor section at an intermediate column;

Fig. 6 is a plan section illustrating the interlocking of columns and wall units;

Fig. 7 is an elevation of column units showing the telescoping element for assembling a plurality of column units;

Fig. 8 is an elevation of a column cap unit;

Fig. 9 is a plan view of a column unit with a subjacent column unit telescoped therein;

Fig. 10 is a plan view of two spaced column units connected together to permit formation of a "stepped back" construction;

Fig. 11 illustrates a stair section and means for interlocking the same with a column; and

Figs. 12, 13 and 14 are plan views showing column sections which permit interlocking of adjacent columns or wall sections either at 45° or 90° angles.

In general the building elements comprise three basic units: interlocking column, wall, and floor units 7, 8 and 9. The wall units come in the form of doors, windows or balustrades and advantageously are of varying widths which are whole multiples of a minimum unit width which, in this case, is that of a single window or door 10 and 11 (Fig. 2).

The column and wall interlocking means is here represented by head and slot elements, the head being embodied in marginal ribs 12 on the wall units and the slots being open ended slots 13 in each of the four column faces (Fig. 6). The ribs are necked down as at 14 under the heads to the width of the slots 13.

The slotted column unit faces are integrally united adjacent one end of each of the column units by a star shaped stud 15 (Fig. 7) which projects substantially beyond one end of the unit and is adapted to enter and

frictionally fit in the interior of a superposed column unit which constitutes a socket for the stud, thereby connecting the column units together with the open ends of the slots 13 in alignment and providing continuous slots in each of the column faces extending from top to bottom of the assembled column. In the present building each star stud 15 is longer than the column section is deep by a distance equivalent to the floor thickness, for purposes presently explained. It spaces the column units by contact with the inner end of the star stud of the adjacent column unit. The spaces between the several flanges 16 of studs 15 are adapted to receive the enlarged head 12 on the wall and door units. Assembly is effected by passing the head 12 into the open end of slot 13 or vice versa.

The column sections are advantageously formed into lengths which are a simple fraction of the height of the wall and door units. In this case the length of each column unit is one-third of the height of the wall units, the wall and door units extending from floor to ceiling. Thus, for each story of the building the tops of the wall and column units are flush and provide a flush surface 18 to receive the margins of the floor sections 9 (Fig. 2).

The material for the floor sections is in sheet form 20 (Fig. 3) and is advantageously subdivided by straight and star shaped weakened or "break" lines 21 and 22 respectively, into variously sized areas 23 whose dimensions are whole multiples of the combined unit width of a wall and column unit, to adapt the same to various arrangements. For example, the floor section 24 embraces the square area enclosed by single unit wall elements, whereas the section 25 is a rectangular area of two units width and four long. The floor blanks may also carry decorative patterns simulating block or tile flooring, carpets, etc.

The star shaped "break-out" pieces correspond with the star shaped sections of the studs 15 of the columns, and when removed leave star shaped recesses in the floor section margins which are complimentary to and register with the star shaped ribs of a column stud whether at a corner or at an intermediate position in a wall (Figs. 4 and 5); and their spacing corresponds to the unit or multiple unit widths of wall or door elements. Thus when a properly proportioned floor section 23 is seated on the aforesaid flush surface 18 of the column and wall tops, the star shaped marginal recesses 22 fit around the registering stud flanges which thereby hold the section against displacement. The superposed column and wall sections for the next story rest on the underlying floor section margins to hold the same down.

The tops of the stair and wall units are provided with spacing pads 26 whose vertical thickness is equivalent to that of the floor unit, thereby to maintain horizontal registry of the wall and stair units throughout the building structure, in the event that a floor unit is omitted in certain portions of the structure to provide multi-storied ceilings.

The surplus length on the column studs 15 serves to space the column units by a distance which simulates a conventional horizontal mortar joint between the units. Columns may be located intermediate corner columns as at 27 (Fig. 6) by interlocking wall and door elements in opposite slots 13 in the column. Such columns as well as corner columns firmly unite the several elements of the structure. The column connecting studs 15 serve also to connect superposed stories together into a substantially rigid structure.

The several structural elements are advantageously die cast from an appropriate thermo-setting plastic such as polystyrene or other tough and slightly flexible yet stiff material, which permits the use of thin sections with close dimensional tolerances and accurate simulation of

3

conventional building design including superficial ornamentation. It is thus possible to provide close sliding friction fits between the interlocking wall and column units to eliminate looseness between the units and thus to achieve substantial rigidity in the assembled structure. As shown in Fig. 6, for example, the interlocking head 12 of a wall or door unit makes line contacts with adjacent flanges of the column studs 15 and the thin section of the flanges permits the slight flexibility necessary to maintain a friction contact sufficient to prevent the units from sliding apart and yet to permit easy interlocking of the heads 12 in slots 13. Similarly, in the case of studs 15, the free ends of the column faces, connected together as they are only at one end, can flex outwardly sufficiently frictionally to grip an inserted stud (Fig. 9).

The sheet floor material may also advantageously be made from a plastic sheet embossed to provide the aforesaid "break" lines. The latter may of course be located to provide floor sections of shapes other than those illustrated in the drawings.

The wall and door sections may not only embody different designs and arrangements but may have varying superficial ornamentation.

Various accessories are advantageously provided to improve the simulation of conventional buildings. For example, the uppermost column studs 15 may be covered with a finishing cap 28 (Fig. 8) having slots 20 therein registering with the column slots 13. The cap may advantageously be located to rise above the level of a floor, and a low balustrade 30 interlocked with the slots in the cap to simulate conventional construction. The top 31 of the cap is advantageously provided with a hole 32 into which a flag pole 33 or a mast may be inserted if desired.

To permit one story to be set back from the next lower story, if desired, offset and joined column sections 23 (Fig. 10) are provided by which "stepped back" columns may be connected to an adjacent column (see Fig. 1). In the present instance the "stepped back" column unit is connected to a column cap to permit the "stepped back" column 34 to rise from the level of a floor (see Fig. 1).

Similarly, as illustrated in Figs. 12, 13 and 14, the column sections may be formed to permit interlocking of adjacent column sections and walls at 45° and 90° angles. Column sections may be provided with ribs 35 (similar to the ribs 12 of the wall units) either along a corner or face (see Figs. 12 and 14 respectively); or a slot 36 (similar in function to the slots 13) at a corner (see Fig. 13), the latter permits the interlocking of a wall unit at an angle of 45° to other units. These figures illustrate some of the possible floor plans thus made possible.

Another accessory is a stair unit 37 (see Fig. 11) having at its sides head elements 38 similar to the head elements 12 on the wall and door units, by which the stair unit may be interlocked with a column slot at any appropriate place in the building. Each stair unit preferably extends a unit distance along a wall so that successive stair units may be connected to successive column slots along a given wall; and the height of each stair unit is a simple fraction (one-third in this case) of the height of a wall unit, so that the top and bottom stair sections will register with the respective floor levels.

It will be observed that the exposed slots in the column faces appear simply as lines in the column faces which harmonize with the superficial design. It will also be apparent that the basic column, wall and floor units permit great flexibility in design and arrangement as well as superficial ornamentation on the units themselves.

Obviously the invention is not limited to the details of the illustrative construction since these may be variously modified. Moreover it is not indispensable that all features of the invention be used conjointly since various features may be used to advantage in different combinations and subcombinations.

4

Having described my invention I claim:

1. A building toy comprising in combination a plurality of telescoped column units of rectangularly arranged faces, each face having longitudinally extended slots therein open at each end and extending continuously from end to end of the unit, each unit having a longitudinally extending recess behind and wider than the slot and also extending from end to end of said unit, each unit having a non-circular stud projecting from one end only and adapted to be telescoped into the open end of an adjacent unit to connect the same together and to register said slots to provide a continuous slot open at both ends extending from top to bottom of the assembled column units, and a wall unit having along a side edge a marginal rib with an enlarged head of a size to fit into a recess behind the slot in a column and a reduced neck below said head adapted to slide in said slot, said wall unit and assembled column being interlocked by a relative longitudinal movement wherein the said head enters the said recess at an open end, said stud connecting said faces together in a unitary structure and holding said faces adjacent said slots substantially rigid against lateral separation whereby the enlarged head of said wall unit is held in said recess against lateral separation from said column unit.

2. A toy building block comprising a hollow column unit of generally square section, a face of said unit having a continuous slot open at both ends extending from top to bottom of the unit, a stud projecting from one end only of said unit and comprising a plurality of vertical ribs with a space between the same, the opposite end of said unit being hollow and adapted to receive a like stud of an adjacent unit, the faces of said column separated by said slot being each connected to a rib of the first named stud into a unitary structure, the space between ribs registering with and being wider than said slot, a wall unit having a rib projecting into said slot and having an enlarged head locked in said space against lateral separation therefrom.

3. A toy building block structure comprising in combination a hollow column unit having a longitudinal slot in its face and a longitudinally extending inner recess behind and larger than said slot and communicating with said slot, a stud of non-circular section projecting from one end only of said unit adapted to enter the opposite hollow end of a like unit to connect the units together with said slots in alignment to form a continuous slot extending from one end to the other of the assembled units, and a wall unit having along a side edge a rib of a thickness corresponding to the width of said slot and a terminal head larger than said slot and adapted to interlock in said recess by telescoping therewith, said stud connecting the portions of said face adjacent said slot together rigidly against lateral separation, thereby preventing release of said head from said recess by lateral movement.

4. A toy building block structure comprising in combination a plurality of alternating column and wall units having a vertically extending rib and slot connecting means to interlock the units against lateral separation, said rib connecting means having a head larger than said slot, and said slot connecting means being constructed and arranged to prevent lateral separation of column and wall units, said rib and slot connecting means being engageable with said slot only by longitudinal telescoping of rib and slot, the tops of the wall and column units being flush with a floor level, each column unit having a stud of non-circular section projecting above the floor level to telescope with a superposed column unit rising above said floor level, and a floor unit resting on the tops of wall and column units at the said floor level and having cutout portions of a non-circular section corresponding to that of said stud and surrounding portions of said stud to provide interlocking means to prevent lateral separation of stud and floor section, and a second level of col-

2,968,118

5

umn units fitting over said studs and wall units interlocked with the latter column units and resting on said floor unit.

5. A toy building block structure comprising in combination a plurality of alternating column and wall units having a vertically extending rib and slot connecting means to interlock the units against lateral separation, said rib connecting means having a head larger than said slot and said slot connecting means being rigid to prevent lateral separation of column and wall units, said rib and slot connecting means being engageable only by longitudinal telescoping of rib and slot, the tops of the wall and column units being flush with a floor level, each column unit having a ribbed stud projecting above the floor level to telescope with a superposed column unit rising above said floor level to align the connecting means on the telescoped column units, and a floor unit resting on the tops of wall and column units at the said floor level, said floor unit having interlocking means including cut-out portions registering with and complementary to and intelocking with the ribbed portion of said studs to interlock the columns with the floor units, and a second level of column units fitting over said studs and wall units interlocked with the latter column units and resting thereon.

6

6. In a toy building structure the combination comprising a plurality of column units each having star-shaped studs projecting from one end and sockets at the other end adapted to receive the star-shaped stud of a like unit, a floor unit comprising a sheet of material having star-shaped recesses therein complementary to and fitting over registering star-shaped studs of column units and thereby interlocking therewith against lateral separation, and other column units fitting over the aforesaid registering studs and resting upon the floor unit.

References Cited in the file of this patent
UNITED STATES PATENTS

311,793	Stranders	Feb. 3, 1885
1,286,462	Wesche	Dec. 3, 1918
1,426,087	Metcalfe	Aug. 15, 1922
1,870,978	Wolfe	Aug. 9, 1932
2,031,848	Ogden	Feb. 25, 1936
2,112,247	McLoughlin	Mar. 29, 1938
2,338,297	Slaughter	Nov. 6, 1945
2,497,657	Cole	Feb. 14, 1950
2,636,312	Martin	Apr. 28, 1953
2,747,325	Kutscha	May 29, 1956
2,800,743	Meehan et al.	July 30, 1957

Patentsammlung Puppenhäuser II • System und Technik • Reihe: Vorlage - Konzept - Konstruktion • Band: 2
Herausgeber: www.atelier-kalai.de, Kerstin Winter • Hersteller / Verlag: Books on Demand GmbH Norderstedt

3

conventional building design including superficial ornamentation. It is thus possible to provide close sliding friction fits between the interlocking wall and column units to eliminate looseness between the units and thus to achieve substantial rigidity in the assembled structure. As shown in Fig. 6, for example, the interlocking head 12 of a wall or door unit makes line contacts with adjacent flanges of the column studs 15 and the thin section of the flanges permits the slight flexibility necessary to maintain a friction contact sufficient to prevent the units from sliding apart and yet to permit easy interlocking of the heads 12 in slots 13. Similarly, in the case of studs 15, the free ends of the column faces, connected together as they are only at one end, can flex outwardly sufficiently frictionally to grip an inserted stud (Fig. 9).

The sheet floor material may also advantageously be made from a plastic sheet embossed to provide the aforesaid "break" lines. The latter may of course be located to provide floor sections of shapes other than those illustrated in the drawings.

The wall and door sections may not only embody different designs and arrangements but may have varying superficial ornamentation.

Various accessories are advantageously provided to improve the simulation of conventional buildings. For example, the uppermost column studs 15 may be covered with a finishing cap 28 (Fig. 8) having slots 20 therein registering with the column slots 13. The cap may advantageously be located to rise above the level of a floor, and a low balustrade 30 interlocked with the slots in the cap to simulate conventional construction. The top 31 of the cap is advantageously provided with a hole 32 into which a flag pole 33 or a mast may be inserted if desired.

To permit one story to be set back from the next lower story, if desired, offset and joined column sections 23 (Fig. 10) are provided by which "stepped back" columns may be connected to an adjacent column (see Fig. 1). In the present instance the "stepped back" column unit is connected to a column cap to permit the "stepped back" column 34 to rise from the level of a floor (see Fig. 1).

Similarly, as illustrated in Figs. 12, 13 and 14, the column sections may be formed to permit interlocking of adjacent column sections and walls at 45° and 90° angles. Column sections may be provided with ribs 35 (similar to the ribs 12 of the wall units) either along a corner or face (see Figs. 12 and 14 respectively); or a slot 36 (similar in function to the slots 13) at a corner (see Fig. 13), the latter permits the interlocking of a wall unit at an angle of 45° to other units. These figures illustrate some of the possible floor plans thus made possible.

Another accessory is a stair unit 37 (see Fig. 11) having at its sides head elements 38 similar to the head elements 12 on the wall and door units, by which the stair unit may be interlocked with a column slot at any appropriate place in the building. Each stair unit preferably extends a unit distance along a wall so that successive stair units may be connected to successive column slots along a given wall; and the height of each stair unit is a simple fraction (one-third in this case) of the height of a wall unit, so that the top and bottom stair sections will register with the respective floor levels.

It will be observed that the exposed slots in the column faces appear simply as lines in the column faces which harmonize with the superficial design. It will also be apparent that the basic column, wall and floor units permit great flexibility in design and arrangement as well as superficial ornamentation on the units themselves.

Obviously the invention is not limited to the details of the illustrative construction since these may be variously modified. Moreover it is not indispensable that all features of the invention be used conjointly since various features may be used to advantage in different combinations and subcombinations.

4

Having described my invention I claim:

1. A building toy comprising in combination a plurality of telescoped column units of rectangularly arranged faces, each face having longitudinally extended slots therein open at each end and extending continuously from end to end of the unit, each unit having a longitudinally extending recess behind and wider than the slot and also extending from end to end of said unit, each unit having a non-circular stud projecting from one end only and adapted to be telescoped into the open end of an adjacent unit to connect the same together and to register said slots to provide a continuous slot open at both ends extending from top to bottom of the assembled column units, and a wall unit having along a side edge a marginal rib with an enlarged head of a size to fit into a recess behind the slot in a column and a reduced neck below said head adapted to slide in said slot, said wall unit and assembled column being interlocked by a relative longitudinal movement wherein the said head enters the said recess at an open end, said stud connecting said faces together in a unitary structure and holding said faces adjacent said slots substantially rigid against lateral separation whereby the enlarged head of said wall unit is held in said recess against lateral separation from said column unit.

2. A toy building block comprising a hollow column unit of generally square section, a face of said unit having a continuous slot open at both ends extending from top to bottom of the unit, a stud projecting from one end only of said unit and comprising a plurality of vertical ribs with a space between the same, the opposite end of said unit being hollow and adapted to receive a like stud of an adjacent unit, the faces of said column separated by said slot being each connected to a rib of the first named stud into a unitary structure, the space between ribs registering with and being wider than said slot, a wall unit having a rib projecting into said slot and having an enlarged head locked in said space against lateral separation therefrom.

3. A toy building block structure comprising in combination a hollow column unit having a longitudinal slot in its face and a longitudinally extending inner recess behind and larger than said slot and communicating with said slot, a stud of non-circular section projecting from one end only of said unit adapted to enter the opposite hollow end of a like unit to connect the units together with said slots in alignment to form a continuous slot extending from one end to the other of the assembled units, and a wall unit having along a side edge a rib of a thickness corresponding to the width of said slot and a terminal head larger than said slot and adapted to interlock in said recess by telescoping therewith, said stud connecting the portions of said face adjacent said slot together rigidly against lateral separation, thereby preventing release of said head from said recess by lateral movement.

4. A toy building block structure comprising in combination a plurality of alternating column and wall units having a vertically extending rib and slot connecting means to interlock the units against lateral separation, said rib connecting means having a head larger than said slot, and said slot connecting means being constructed and arranged to prevent lateral separation of column and wall units, said rib and slot connecting means being engageable with said slot only by longitudinal telescoping of rib and slot, the tops of the wall and column units being flush with a floor level, each column unit having a stud of non-circular section projecting above the floor level to telescope with a superposed column unit rising above said floor level, and a floor unit resting on the tops of wall and column units at the said floor level and having cut-out portions of a non-circular section corresponding to that of said stud and surrounding portions of said stud to provide interlocking means to prevent lateral separation of stud and floor section, and a second level of col-

UNITED STATES PATENT OFFICE.

RAY H. GROVES, OF PORTLAND, OREGON, ASSIGNOR OF TWENTY-EIGHT ONE-HUNDREDTHS TO E. J. CLOUGH AND THIRTY-SIX ONE-HUNDREDTHS TO GLENN W. PERCIVAL, BOTH OF PORTLAND, OREGON.

INTERLOCKING BUILDING-BLOCK.

1,271,160.			Specification of Letters Patent.			Patented July 2, 1918.

Application filed November 5, 1917.   Serial No. 200,362.

To all whom it may concern:

Be it known that I, RAY H. GROVES, a citizen of the United States, residing at Portland, in the county of Multnomah and State of Oregon, have invented a new and useful Interlocking Building-Block, of which the following is a specification.

My invention relates to improvements in interlocking building blocks for toy structures, and the objects of my invention are to provide blocks which will permit the construction of a large variety of structures of pleasing appearance and of sufficient rigidity to allow handling without danger of collapse.

I attain these objects by the means illustrated in the accompanying drawing in which—

Figure I shows my blocks assembled into a structure, in this case a house, Figs. II to XIV represent a limited variety of individual pieces to illustrate the principle on which my invention is based.

Similar numerals refer to similar parts throughout the several views.

The material employed in the manufacture of my blocks is, preferably, wood although any other suitable substance may be used.

In thickness all the members are equal but in width they conform to a certain system, while the length varies according to requirements, as do also the number and disposition of notches in the individual members.

According to aforementioned system my blocks are divided into three groups.

The first group comprises members of single width having one or more square notches on one longitudinal edge only. Figs. II, III and IV represent samples of this group. They are chiefly used as starting and finishing members for vertical walls, columns, chimneys, etc.

The second group comprises members of double width; their length and the number and location of notches vary according to requirements, specimen being shown in the drawing under Figs. V, VI and VII. Members of this group form the major portion of vertical walls, built-up columns, chimneys, etc. The pieces designated as Figs. VIII, IX and X also belong to this group, though being used in the construction of gables, one or both of their ends are cut slantingly to conform to the pitch of the roof, and the direction of the notches adjacent to these ends is parallel to the slant.

Figs. XI and XII represent pieces of triple width and belonging to the third group; they are used chiefly to form roofs or other slanting surfaces. Here again various lengths may be employed and each longitudinal edge may have two or more notches according to requirements.

Fig. XIII requires special notice. It is a short member of the second group but the two notches, located opposite each other, are of double width to receive two members side by side; as for instance in connecting pieces like Fig. IV into a continuous fence, barrier or similar structure with Fig. XIII representing the supporting post.

The piece in Fig. XIV also has a special function, and that is, to interlock two members, one above the other, only one of which is provided with a notch. Fig. XIV being in width midway between group one and two, its center portion 1, when dropped into notch 2 of Fig. VII for example, will not project above edge 3, so that a piece like Fig. IV with its edge 4 can join closely to edge 3, yet will be prevented from lateral motion, at least, by projections 5 and 6 of Fig. XIV.

Referring now to Fig. I, the use of my building blocks will be clearly understood. At point 7 a suitable member of the first group is laid on the floor; at 8, with end notches interlocking, a suitable member of the second group is laid across the former; across the latter again a member of the second group is laid, and so on until the desired height is obtained, when a member of the first group may be used to level off the structure. Where doors 9 or windows 10, 11 or 12 are desired, pieces Fig. V may be employed to interlock the ends abutting the openings. At the point 13 use is made of Fig. VIII. Point 14 shows the application of Fig. X. Numerals 15 and 16 indicate where members of the third group are used as roofs. It will be noted that the notches in members of the third group are set back farther from the end than in members of the first and second group. This is done purposely to give the roofs the customary overhanging appearance.

To show the adaptability of various members to various purposes it may be pointed out that two longer members 17 and 18 of the first group with a shorter member 19 interlocked at the top make very suitable porch columns and roof supports.

Having thus described my invention it will be seen that my objects have been accomplished; and, though it is not practicable to show all the possible variations in blocks and structures, I reserve to myself the right to change minor details without violating the spirit of my invention.

I claim:

1. In interlocking building blocks of the character described, the combination of single, double and triple width elements into a toy structure designated as Fig. I, the single and double width elements forming the walls, gables, columns and chimneys, the triple width elements forming sloping roofs, said roofs having a pronounced overhang at their gable ends, substantially as described.

2. As an article of manufacture, interlocking building blocks of equal thickness but various lengths, having a width approximately four times their thickness, having one or more square notches either on one or both of their longitudinal edges, and having either one or both of their ends cut slantingly, with the direction of the notches adjacent to the slant being parallel to said slant.

3. As an article of manufacture, an interlocking building block, having two square notches opposite each other and the width of each notch equaling twice the thickness of the block; said block being designated as Fig. XIII.

4. As an article of manufacture, an interlocking building block, having a width approximately three times, and a length approximately four times its thickness, having two square notches opposite each other, and designated as Fig. XIV.

RAY H. GROVES.

United States Patent [19]

Ahlstrand

[11] **3,751,848**

[45] **Aug. 14, 1973**

[54] **MODEL HOUSE**

[76] Inventor: **Richard G. Ahlstrand**, 1281 Hearst Ave., Berkeley, Calif. 94702

[22] Filed: **Mar. 13, 1972**

[21] Appl. No.: **233,904**

[52] **U.S. Cl.** **46/19, 46/21, 35/16**
[51] **Int. Cl.** **A63h 33/06**
[58] **Field of Search** 46/12, 19, 21; 35/16

[56] **References Cited**

UNITED STATES PATENTS

1,867,374	7/1932	Myers	46/21
1,891,011	12/1932	Purdy	46/21
3,234,680	2/1966	Starr	46/21

Primary Examiner—Louis G. Mancene
Assistant Examiner—Robert F. Cutting
Attorney—Roger W. Erickson

[57] **ABSTRACT**

A model house that can be erected from a combination of parts without tools or fasteners and then easily knocked down for shipment or storage. A base member has a main cutout to receive a central vertical partition and peripheral openings to receive vertical side members that support a second floor and a folding roof. All parts are rigid members that interlock so that they cooperate together to maintain the structural arrangement of the house.

14 Claims, 6 Drawing Figures

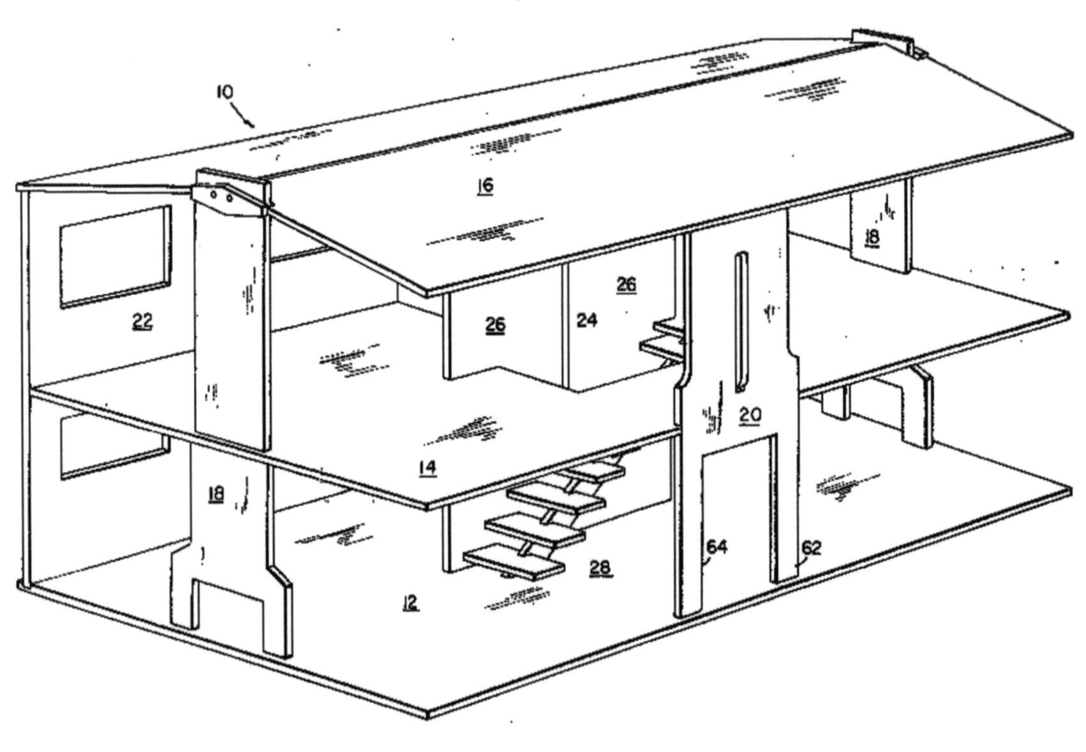

Patentsammlung Puppenhäuser II • System und Technik • Reihe: Vorlage - Konzept - Konstruktion • Band: 2
Herausgeber: www.atelier-kalai.de, Kerstin Winter • Hersteller / Verlag: Books on Demand GmbH Norderstedt

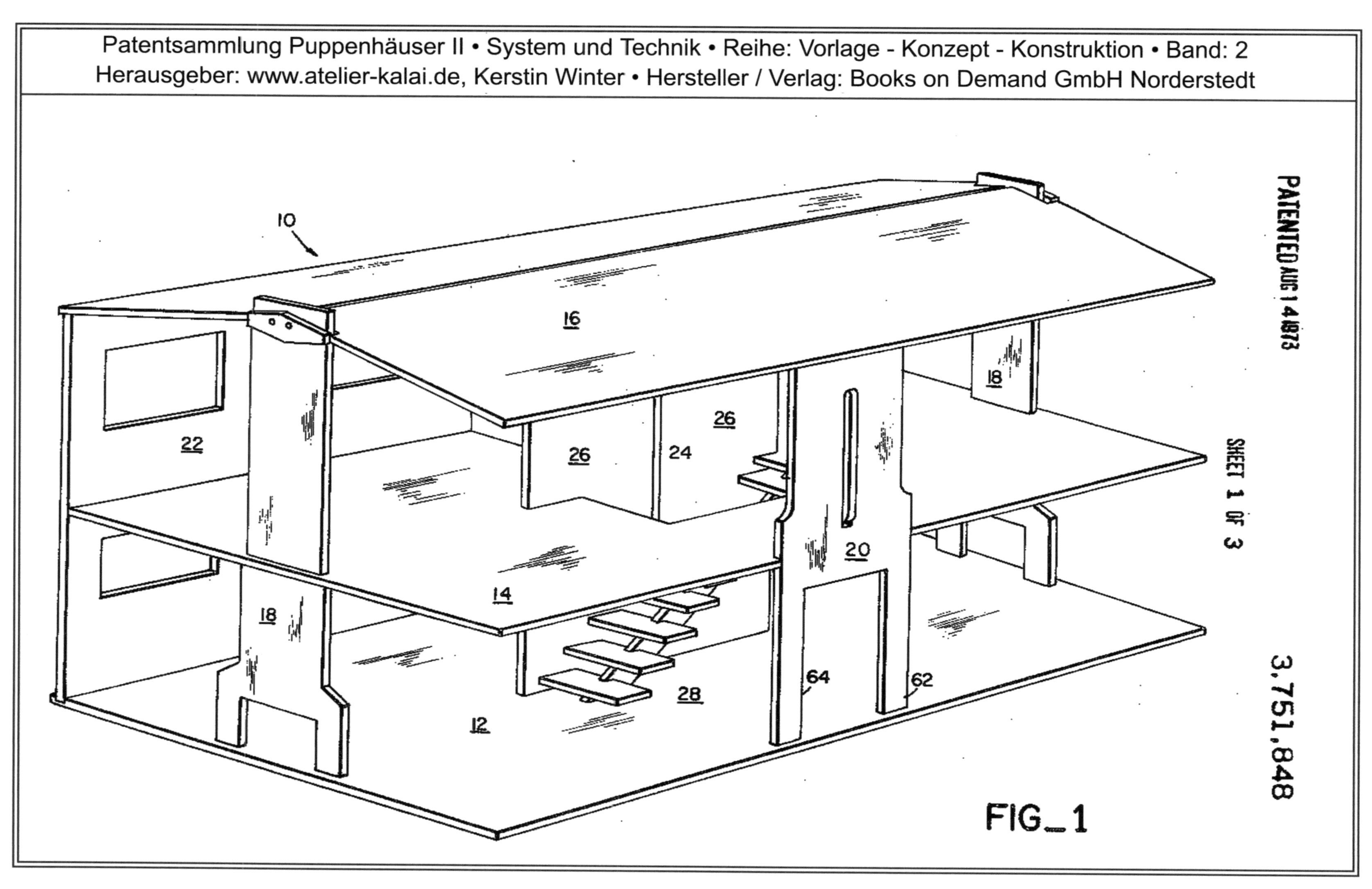
PATENTED AUG 1 4 1873
3,751,848
SHEET 1 OF 3
10
16
18
22
26
26
24
26
20
14
18
28
64
62
12
FIG_1

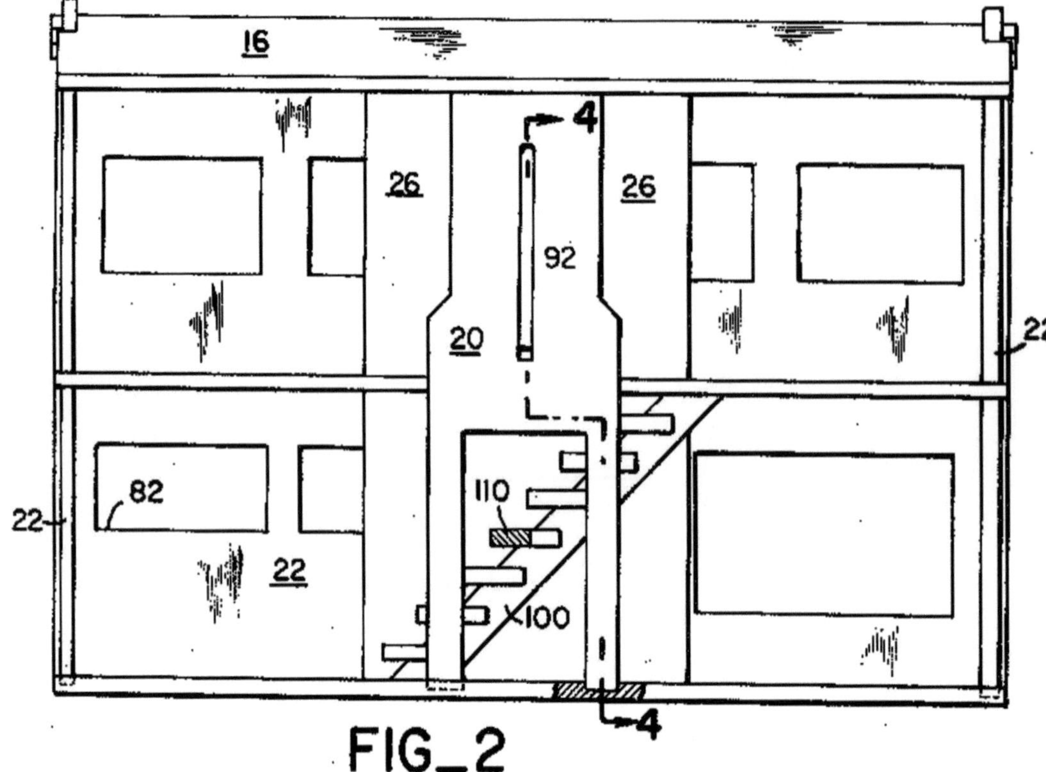

16
26
26
92
20
22
22
82
22
110
100
4
4
FIG_2

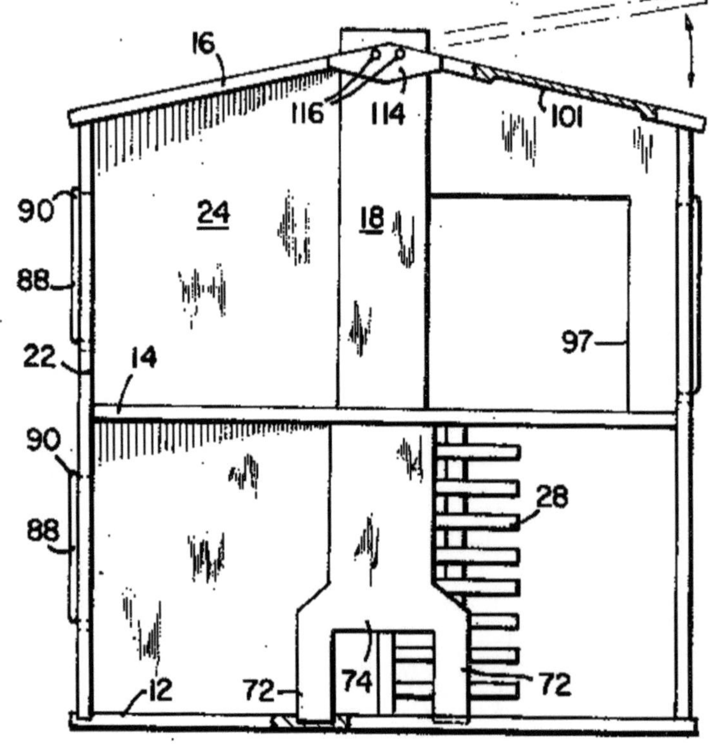

16
116 114
101
90
24
18
88
97
22
14
90
28
88
12
72
74
72
FIG_3

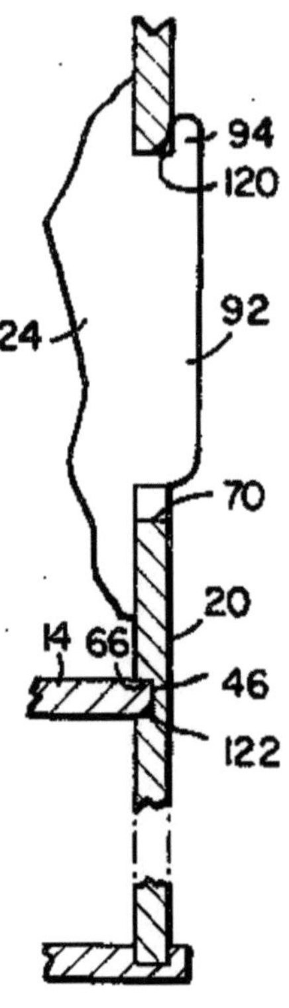

94
120
24
92
70
20
14 66
46
122
FIG_4

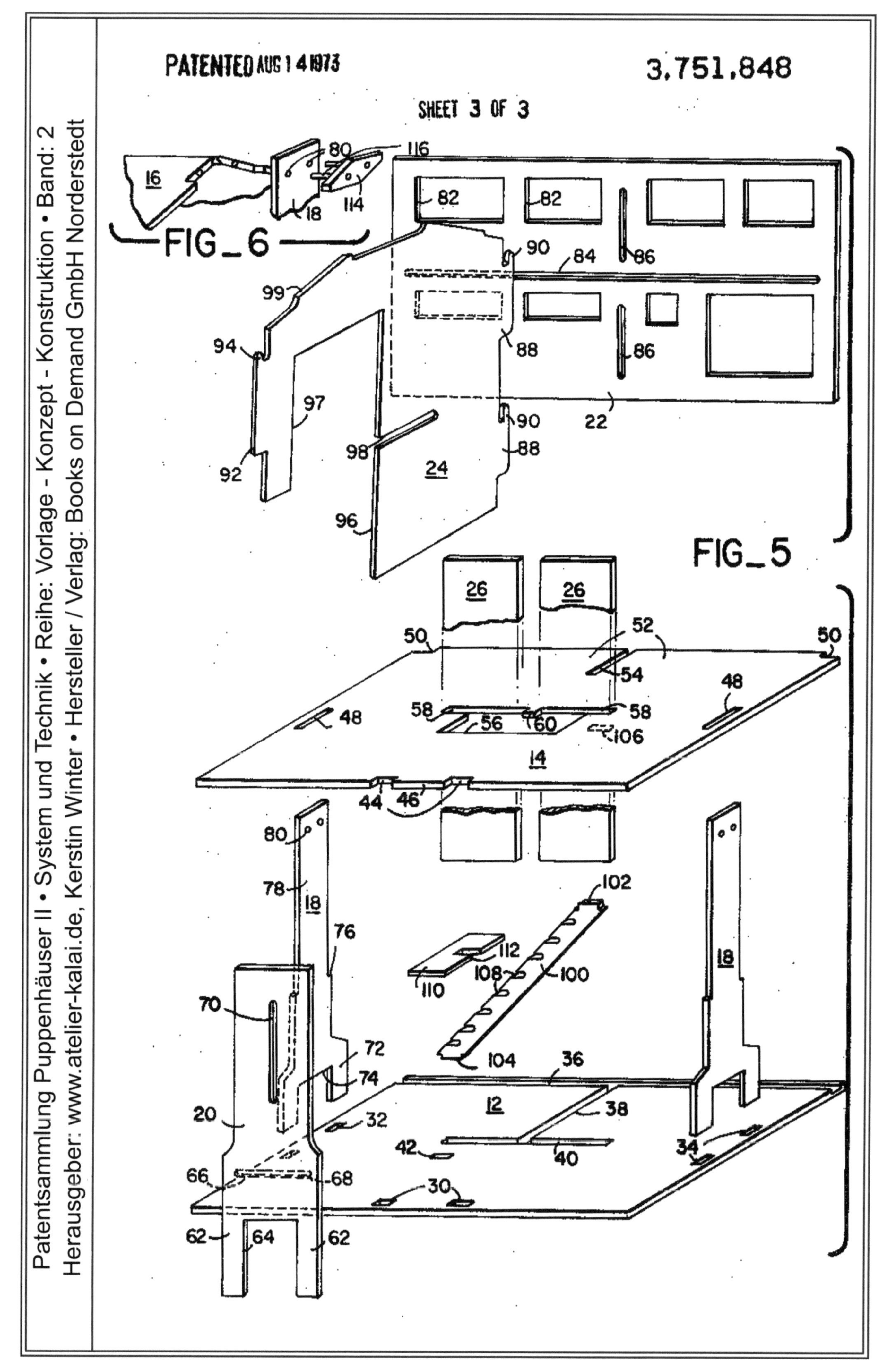

PATENTED AUG 14 1973
3,751,848
SHEET 3 OF 3
FIG_6
16
80
116
114
18
FIG_5
99
94
97
92
98
24
96
82
82
90
84
86
88
86
22
90
88
26
26
50
52
54
50
48
58
56
60
58
106
48
14
44
46
80
78
18
102
76
112
108
110
100
70
18
72
74
104
36
20
12
32
38
42
40
34
66
68
30
62
64
62

1

MODEL HOUSE

BACKGROUND OF THE INVENTION

This invention relates to a model house structure which is useful as a means for displaying or planning interior designs or as a child's doll house.

Heretofore doll houses of various types have been deviced to simulate actual home interiors and thereby provide a recreational and educational structure for children. For the most part such houses were generally permanent structures constructed in a conventional manner with numerous parts and fasteners. Once constructed or assembled these houses were not adaptable to be easily disassembled for storage or movement to another locality.

It is therefore one object of the present invention to provide a doll house that is comprised of relatively few parts which can be knocked down for shipment and storage and yet can be easily created for use without the need for skilled labor, special tools or fasteners.

Another object of the present invention is to provide a doll house that is realistic and eye-pleasing in appearance.

Yet another object of my invention is to provide a model house that is sturdy, durable and can withstand rigorous use.

Another object of the present invention is to provide a doll house having internal first and second floor areas that are easily accessible from three or more sides of the structure.

Another object of the present invention is to provide a doll house that is well adapted for general ease and economy of manufacture.

Other objects, advantages and features of my invention will become apparent from the following detailed description of one preferred embodiment presented with the accompanying drawings.

BRIEF SUMMARY OF THE INVENTION

In general, my model or doll house comprises a plurality of basic components that form the floors, the roof and side supporting members that may simulate sidewalls but which leave substantial openings to provide considerable access to the interior floor areas from at least three sides of the house. A first floor or horizontal base member has a series of centrally located slots which receive the lower ends of a vertical transverse divider support member. At opposite ends of the base member are holes or slots for receiving the lower ends of narrow, upright support members that may simulate chimneys.

Along the rear edge of the base member is another slot for receiving a rear upright member that may be long enough to simulate a wall and which is connected to the transverse divider member. Yet another narrow, upright member may be secured in openings along or near to the front edge of the base member to receive and retain a front upright member. The transverse divider member together with the front, rear and end support members hold in place a second floor member and a roof member, all of which are locked together when the house is assembled.

BRIEF DESCRIPTION OF THE DRAWINGS

FIG. 1 is a view in perspective of a model house embodying the principles of my invention as it appears when fully assembled;

2

FIG. 2 is a somewhat smaller front view in elevation of the house shown in FIG. 1;

FIG. 3 is an end view in elevation of the house shown in FIG. 1;

FIG. 4 is a fragmentary view in section taken along line 4—4 of FIG. 2;

FIG. 5 is an exploded view in perspective showing all of the major parts of my house except the roof as they appear when separated from each other; and

FIG. 6 is an enlarged fragmentary view in perspective showing one roof connection.

DETAILED DESCRIPTION OF PREFERRED EMBODIMENT

Referring to the drawing, FIG. 1 shows a model house 10 embodying the principles of the present invention as it appears when fully assembled for use as a display or recreational device. Generally, it comprises a base or first floor member 12 having a rectangular shape with a similarly shaped second floor member 14 spaced vertically above it. A hinged two-piece roof member 16 is spaced above the second floor member. Holding the floor and roof members firmly together are a pair of end members 18, a front member 20 and a rear wall member 22, all of which are joined to or engaged with edge portions or near-edge portions of both floor members and the roof member. Centrally located within the house is a transverse, upright divider member 24 which provides support to the floor and roof members and also serves to divide the interior floor into room areas. Perpendicular to this divider are a pair of additional divider members 26. A stairway 28 located adjacent the divider members 26 provides a central access between the first and second floors. Since the end members and the front members can be made relatively narrow in width the interior of the house on both floors is readily accessible. This is an important feature because it greatly enhances the use and enjoyment of the house by children as a doll house.

All of the components of my house 10 are preferably made from a suitable sheet material such as wood or plastic board that is relatively light yet durable and similar in appearance to materials used on actual houses. A fairly thick (e.g., one-half inch) plywood such as birch is particularly adaptable because it can be sanded and finished to have smooth, precisely dimensioned surfaces. Another important feature of my invention is that all components of my house can be made from the aforesaid sheet material so that when disassembled they can all be stacked and packaged together in a relatively compact space for storage or shipment.

Turning to FIGS 2 – 6, the house 10 is shown in greater detail to illustrate the detailed shape and function of various components. The base member 12 has a pair of spaced apart recesses 30 centrally located near its front edge. Similarly spaced apart recesses 32 and 34 are provided at the opposite ends of the base member. Extending parallel to the rear edge of the base member and along its entire length is an elongated groove 36. Perpendicular to this latter groove is a T-shaped groove 38 having an inner portion 40 that is parallel to the rear edge groove 36 and located substantially midway between the front and rear edges of the base member. Near one end of this inner groove portion 40 and spaced from it is a relatively short recess 42. All of the aforesaid grooves and recesses may be routed to a suitable depth within the base member by

3

conventional wood masking tools and it is preferred that none of them extend completely through the base member so that its under surface will remain smooth and unbroken.

The second floor member 14 which has the same rectangular shape as the base member 12 also has cut-out portions generally above its various grooves and recesses. Along the front edge of the base member are a pair of notches 44 between which is formed a tongue member 46. At its opposite ends are a pair of elongated rectangular holes or slots 48. The rear corners are each provided with a notch 50 between which are formed a pair of elongated tongue members 52. Extending inwardly from the rear edge of the floor member 14 between its tongue members 52 is a slot 54 which is vertically aligned with the slot 38 when the house is assembled. Spaced from the end of this slot 54 is an opening 56 having slot portions 58 at opposite ends and along one side thereof. These slot portions are vertically aligned with the groove 40 on the base member when the house is assembled. Separating the slot portions 58 is a short tongue member 60.

At the front of the house the front member 20 has a bifurcated low portion with two spaced apart vertical portions 62 forming an opening 64 that readily simulates a front door. Spaced above this opening on the inside of the front member 20 is a horizontal groove 66 having substantially the same width as the tongue 46. The width of material designated by the numeral 68 at the opposite ends of the groove 66 is the same or slightly less than the width of the slots 44. Spaced above the groove 66 is a vertical slot 70 which cooperates with the upright divider member 24 when the house is erected.

At the opposite ends of the house the end members 18 may be relatively narrow, as shown, so as to simulate fireplaces but also to provide easy access to the interiors of both floors. The low end of each member 18 is bifurcated with narrow spaced apart portions 72 forming an opening 74. Above the portions 72 each member 18 tapers to a narrower width which extends upwardly to the second floor level. Here, end member 18 tapers abruptly to a still narrower width by means of a pair of horizontal shoulders 76 on its opposite sides. The narrowest portion 78 above these shoulders has a width only slightly smaller than the length of the slots 48. Near the upper end of this narrow portion are a pair of spaced apart holes 80.

The rear wall member 22 is generally rectangular and has a series of openings 82 which simulate windows. Its thickness is only slightly less than the width of the groove 36 in the base member. Spaced approximately midway between its upper and lower edges is a horizontal groove 84, the opposite ends of which terminate a short distance from the vertical end edges of the wall member. This groove is just slightly larger in width than the tongue portions 52 of the floor member 14 so as to receive them with a snug fit. Located approximately midway between the ends of the wall member 22 are a pair of vertically aligned and spaced apart slots 86 adapted to receive and retain the divider member 24. This rear wall member, as shown, extends the entire length of the base member and thereby simulates a full outside wall of the house. However, it is apparent that it could be shorter, if desired, particularly if more access to the interior of the house was required from its rear side.

4

The transverse divider member 24 extends between the rear wall 22 and the front member 20 and helps to hold the second floor in place as well to support the roof 16. Along a rear vertical edge of this divider are a pair of projecting tongue members 88 which are spaced apart and sized so as to fit within the slots 86. Each of these members 88 is generally hook shaped with an upper end portion 90 that extends outwardly and upwardly from the vertical edge of the divider. When these tongue members are inserted into the slots 86, as shown in FIG. 4, their end portions 90 extend above the upper ends of the slots and hold the two adjoining members 24 and 26 firmly together. On the opposite vertical edge of the divider 24 is a similar projecting tongue member 92 also with a hook-like upper portion 94. This tongue member fits into the vertical slot 70 of the front upright member 20 in the same manner as the other tongue members 88. The divider 24 is shaped so that it has an internal vertical edge 96 on one side of cutout portion 97 above the stairwell opening 56 and an internal horizontal slot 98 that extends inwardly therefrom towards its rear edge. The length of this latter slot is such that when the divider 24 is attached to the floor member 14 so that its internal edge 96 is flush with the projection 60, the portion of the divider 24 adjacent its slot 98 will fit in the slot 54 of the floor member. At the top of the divider are a pair of tongue members 99 that are sloped at an angle in opposite directions and are adapted to fit within grooves 101 of the same length in the underside of the roof members 16 as shown in FIG. 3.

The other vertical dividers 26 fit into the slots 58 and adjacent to these divers the opening 56 provides space for accommodating the stairway 28. The stairway is comprised of an inclined member 100 with projecting tongue members 102 and 104 at its upper and lower ends. The upper tongue member fits into a recess 106 provided in the underside of the floor member 14 and the lower tongue member fits into the similar recess 42. A series of horizontal slots 108 are provided in one side of the inclined member 100 for receiving a series of steps 110 each having a similar slot 112.

The roof 16 is comprised of two generally rectangular members that are pivotable around a hinge axis that extends parallel to the rear wall member 22. As shown in FIG. 6, both roof members at both ends of the hinge axis are notched out to receive the upper ends of the upright end members 18. At this point each end member is secured to the roof by a short retainer block 114 which extends across the width of the end member and has a pair of spaced apart dowel pins 116 that project from one side of the block. The dowel pin of each block extends through the openings 80 in the end member and into similar openings provided in the roof members. Thus, the dowel pins serve as hinge means by allowing each roof member to pivot about the aligned pins at its opposite ends. When installed in the block the dowel pins are angled outwardly with respect to each other a small amount (e.g., 1° to 2°) and this assures that once placed in their proper position they will essentially grip the sides of their openings and stay in place.

The assembly of my house 10 can be accomplished quickly and easily by an unskilled person, and in fact part of the unique recreational value of the house to small children lies in the process of assembling and disassembling the various components. With the base

5

member 12 supported on a flat surface the center divider 24 is first attached to the rear wall 22 by inserting the rear tongue members 88 into the slots 86. Now, with the back wall preferably resting on a flat surface, the floor member 14 is inserted into the slot 98 of the divider 24 so that its slot 54 receives the divider portion adjacent thereto and until the tongue members 52 of the floor are seated in the groove 84 of the rear wall. The slot 98 of the divider 24 and the groove 84 of the rear wall will be in the same plane to allow connection of these components only when the divider tongues 88 are uppermost in the slots 86. Therefore when the floor member 14 is inserted in the divider 14 and into the back member, the back member becomes locked to the divider. Now, the front upright member 20 is secured in place by inserting the front tongue member 92 on the divider 24, into the slot 70, while the tongue member 46 of the floor 14 fits within the recess or groove 66. This action locks the front member to the divider and in fact locks these members to the floor member 14 in the same manner that it was previously locked to the rear wall member. This connection is made possible by rounding or beveling the top outer edge 120 of the slot 70 and the bottom edge 122 of the floor tongue portion 46. As the front member is connected the rounded edge 120 engages upper portion 94 of the tongue 92 and cams the front member downwardly until the tongue 92 is uppermost in the slot 70 and the tongue portion 46 of the floor member is with the mating slot 66. The end members 18 can now be installed in a similar manner by being inserted upwardly through the floor slots 48. At this point, the assembled components are pivoted into place on the base member with the rear wall situated in the groove 36, the front door member in the recesses 30 and the end chimney members in their recesses 34. Now, the steps 110 can be inserted into the notches 108 of the member 100 to form the stairway 28 and its bottom tongue 104 can be placed in the recess 42 of the base member. By lifting the floor member 14 slightly, the upper tongue member 102 of the stairway can be inserted into the recess 106.

The room dividers 26 can now be inserted into the slots 58 adjacent the opening 56 that forms the stairwell.

Now, the roof 16 is attached to the center divider 24 so that the tongue members 99 thereof are inserted into the grooves 101 of the roof. In this position the upper ends of the chimney like end members 18 are situated within the notched out portions at the opposite ends of the roof members. The retaining blocks 114 can now be attached with their dowel pins 116 inserted through the holes 80 and into the roof.

The disassembly of the house is accomplished by merely reversing the assembly procedure and should be readily apparent. When assembled the house is sturdy and stable with the components locked together as described. Yet no fasteners such as screws or bolts are required. The high degree of accessibility of the floor areas greatly enhances the use and enjoyment of the house by more than one child at a time. In the knocked down condition all parts are flat for easy storage and handling.

To those skilled in the art to which this invention relates, many changes in construction and widely differing embodiments and applications of the invention will suggest themselves without departing from the spirit and scope of the invention. The disclosures and the de-

6

scription herein are purely illustrative and are not intended to be in any sense limiting.

I claim:

1. A model house that can be rapidly assembled from a combination of individual, connectable components without fasteners and thereafter disassembled for storage or shipment, comprising:

a generally rectangular base member providing a first floor means and having a groove extending parallel to one side thereof;

a second generally rectangular member providing a second floor means parallel to said first floor means;

a rear member simulating a rear wall having a generally rectangular shape and extending vertically upward from said groove and substantially the full length of said base member;

a front member having a shorter length than said rear member, supported on and extending vertically upward from said base member and simulating a front portion of the house;

a pair of end vertical members simulating end wall portions of the house located near end edges of said base member and extending vertically upward therefrom, said rear, front and end vertical members all being connected with said first and second floor means;

a transverse divider member having opposite end portions connected with said rear and front members and extending vertically upward from said base member;

and roof means connected to said divider member and said end members.

2. The model house as described in claim 1 wherein said divider member has tongue portions projecting from its vertical edges at its opposite ends which extend through aligned slots in said front and rear vertical members and sloped tongue members that fit within mating grooves of said roof means.

3. The model houses as described in claim 1 wherein each of said end vertical members has an enlarged lower portion and a narrower upper portion forming shoulder means for supporting said second floor means.

4. The model house as described in claim 1 wherein said front vertical member has a bifurcated lower end forming an opening that simulates a front door and a slot on its inner surface for receiving an edge portion of said second floor means.

5. The model house as described in claim 1 including retainer block means having projecting pins that extend through the upper end of each end member and into said roof means.

6. The model house as described in claim 1 wherein said divider member and said second rectangular member have similar slots for connecting them in a cruciform manner, the slotted portion of said divider member having a width that is approximately one half of the width of the second rectangular member, said slotted portion extending from the rear edge thereof so that an open space extending across the entire length of the second floor means is provided.

7. The model house as described in claim 6 including a stairway means extending below a stairwell opening in said open space between said first and second floor means.

8. The model house as described in claim 7 wherein said stairway means comprises an inclined beam mem-

7

ber having a series of spaced apart slots with a step member in each slot and tongue members at the ends of said beam member that fit within recesses of said first and second floor means.

9. The model house as described in claim **8** wherein all of said components are made from sheet plywood material.

10. The model house as described in claim **8** including a pair of internal wall members forming a sidewall of said stairwell opening and situated perpendicular to said divider member.

11. A model house comprising:

a rear upright wall member having a horizontal groove and a pair of vertical slots;

a front upright member spaced apart and parallel to said rear member and having a horizontal groove and a vertical slot;

a vertical divider member extending transversely between said rear and front members, said divider member having a horizontal slot and including locking means projecting from its opposite ends connected to said rear and front members;

a horizontal floor member having opposite edge por-

8

tions which are retained by said grooves in said rear and front members and said divider slot only when said divider locking means are in their locked position within said rear and front members.

12. The model house as described in claim **11** wherein said locking means on said divider member are projecting tongue members extending through said slots in said rear wall member and said front member, said slot on said front member having a rounded outer edge to facilitate its attachment to said divider and said floor member.

13. The model house as described in claim **11** including a base member having recesses for receiving the lower ends of said rear member, front member and divider member.

14. The model houses as described in claim **13** including upright end support members supported on said base member each extending upwardly through an opening in said floor member and having a pair of spaced apart holes at its upper end; pin means extending through said holes and a pair of roof members pivotally supported on said pin means.

* * * * *

United States Patent [19]

Abrams

[11] **4,107,869**

[45] **Aug. 22, 1978**

[54] **DEMOUNTABLE TOY HOUSE**

[75] Inventor: **D. David Abrams**, Kowloon, Hong Kong

[73] Assignee: **Mego Corp.**, New York, N.Y.

[21] Appl. No.: **777,850**

[22] Filed: **Mar. 15, 1977**

[51] Int. Cl.² A63H 33/08
[52] U.S. Cl. 46/21
[58] Field of Search 46/19, 21

[56] **References Cited**

U.S. PATENT DOCUMENTS

1,241,594	10/1917	Williams	46/21
1,428,405	9/1922	Wegener	46/21
1,551,666	9/1925	Jensen et al.	46/21

Primary Examiner—F. Barry Shay
Attorney, Agent, or Firm—Bertram Frank

[57] **ABSTRACT**

A toy house, such as a toy country store, formed by a plurality of interlocking panels which do not rely on separate pins, clips or the like for attachment together. In one embodiment the toy house is formed, without exterior walls, of panels which interlock by means of slots. Another embodiment relies on plastic fittings to hold edges together, and in addition bendable joints between panel sections are provided. The panel members in any case have slits or slots and/or joining edges so that the panels fit together to yield a toy house such as a toy store, or other structures.

8 Claims, 26 Drawing Figures

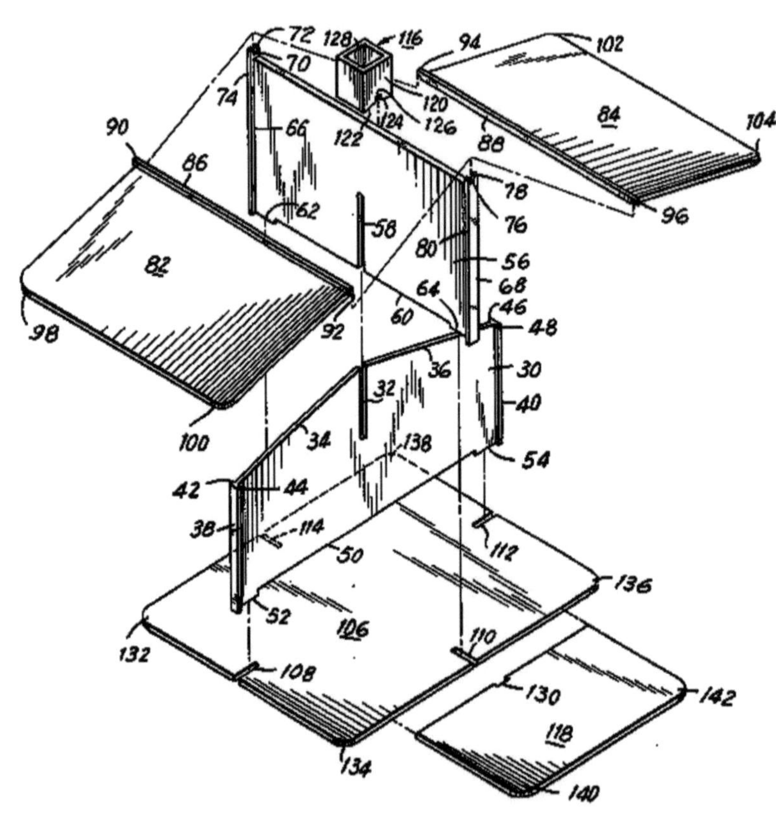

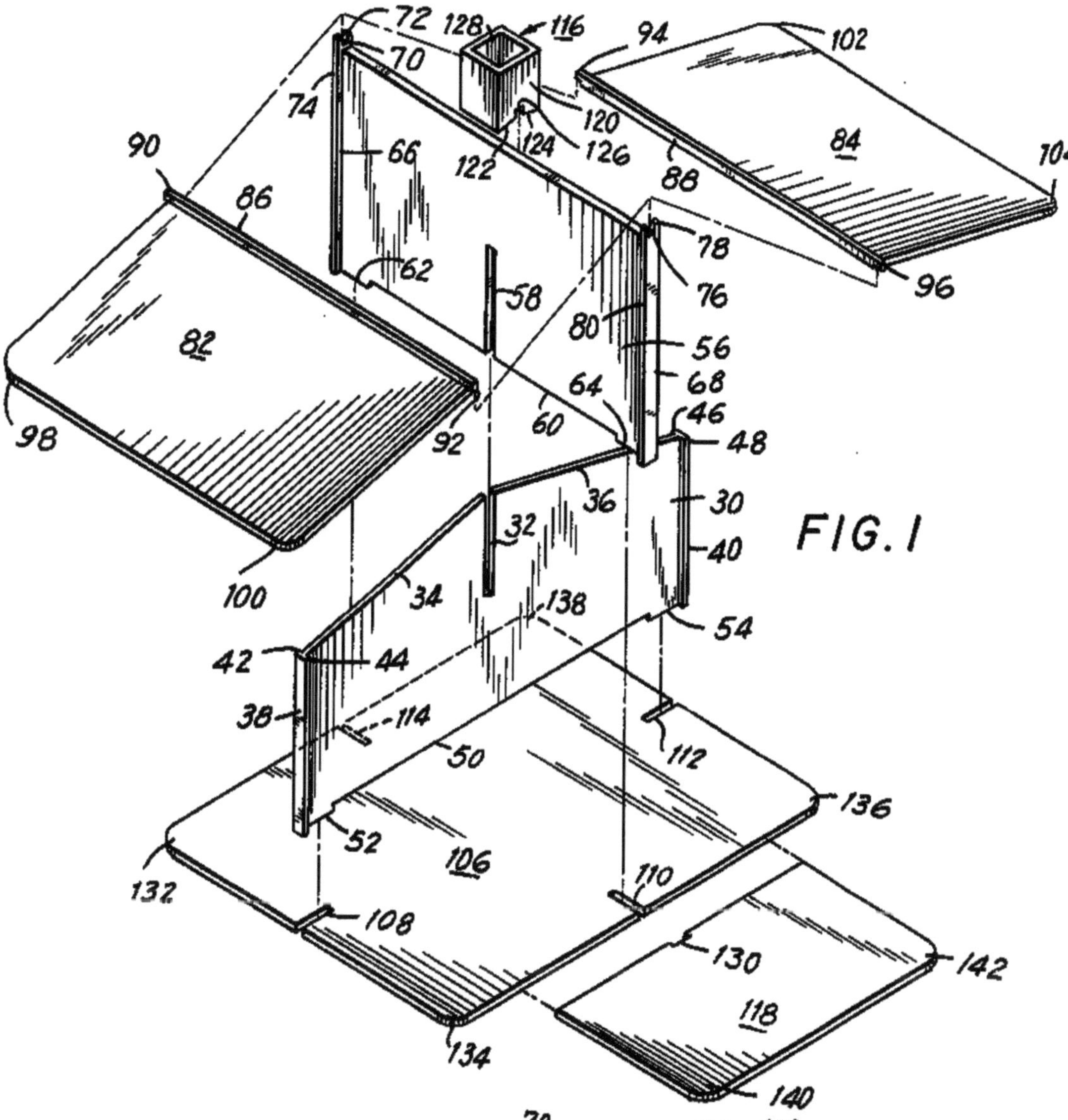
72
128
116
70
74
94
102
120
124
122
126
66
84
104
90
86
88
62
78
58
76
96
80
82
56
68
64
46
98
92
60
48
36
30
32
40
100
34
138
54
42
44
38
114
112
50
136
52
106
110
132
108
130
142
118
134
140

FIG. 1

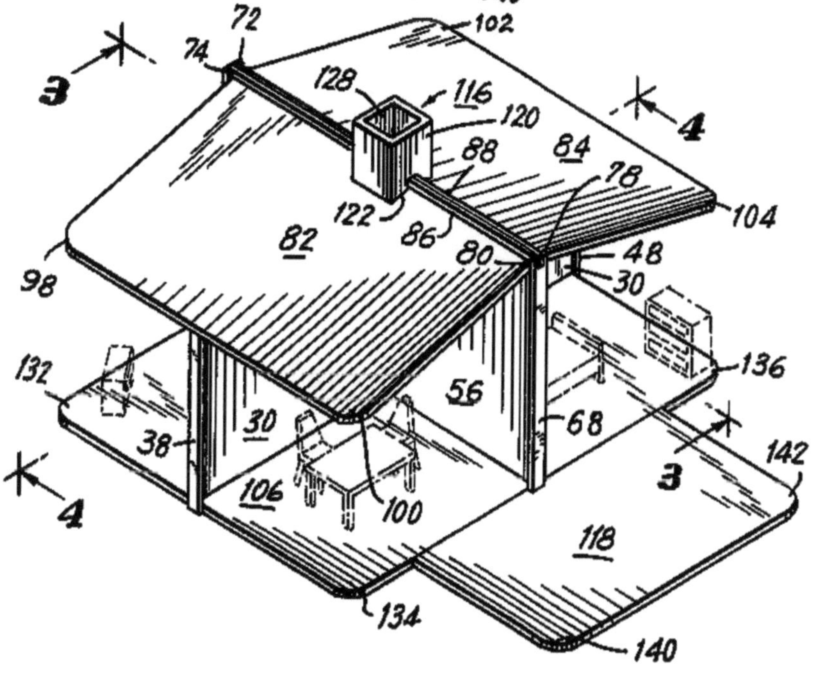
72
102
74
3
128
116
120
4
88
84
78
122
86
104
82
80
48
30
98
136
132
56
30
68
38
100
106
4
3
142
118
134
140

FIG. 2

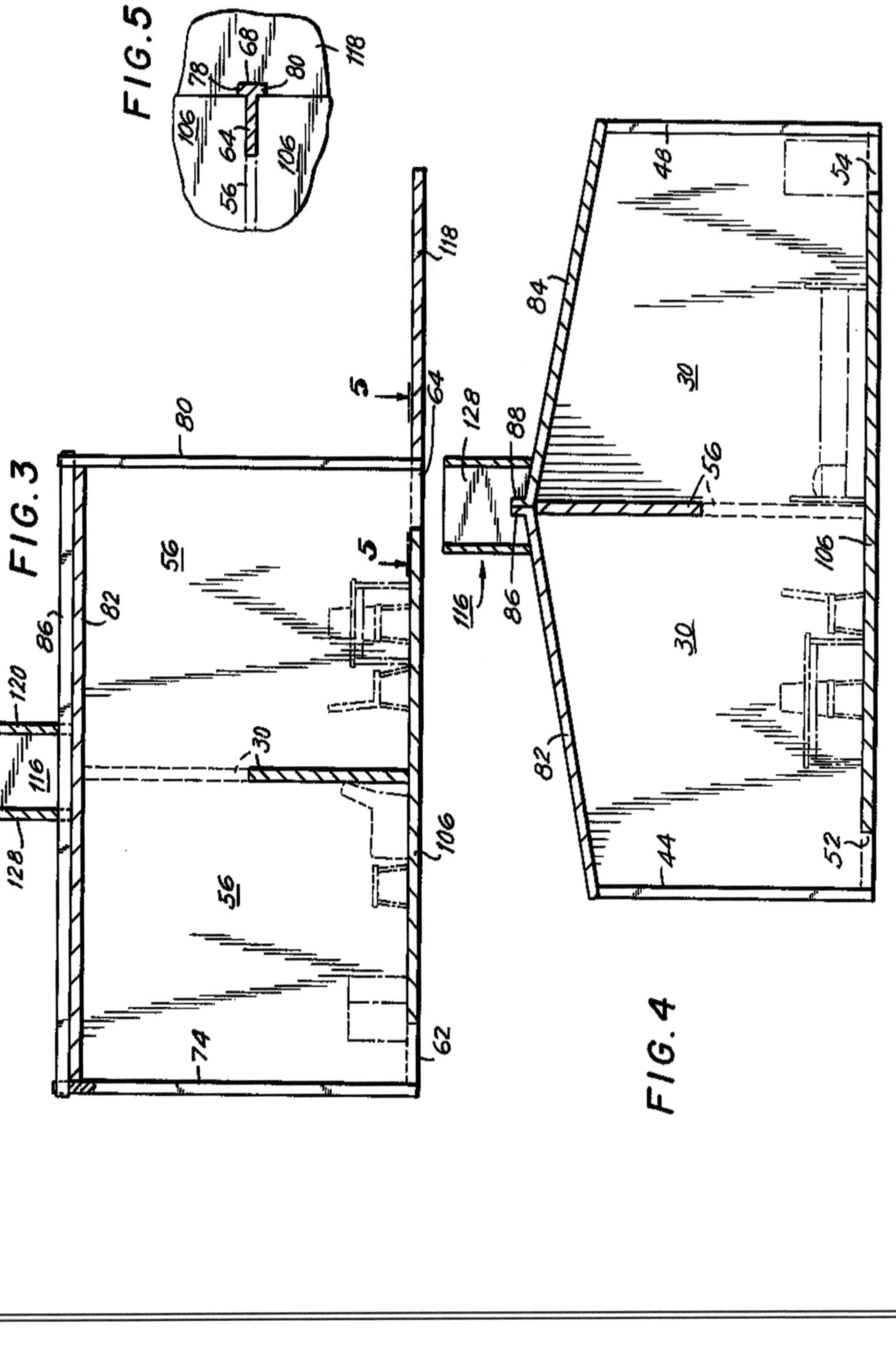
FIG. 5
FIG. 3
FIG. 4
106
78
68
80
118
64
56
118
80
5
64
128
88
84
48
54
30
56
86
116
82
120
86
116
128
30
56
106
74
62
44
52

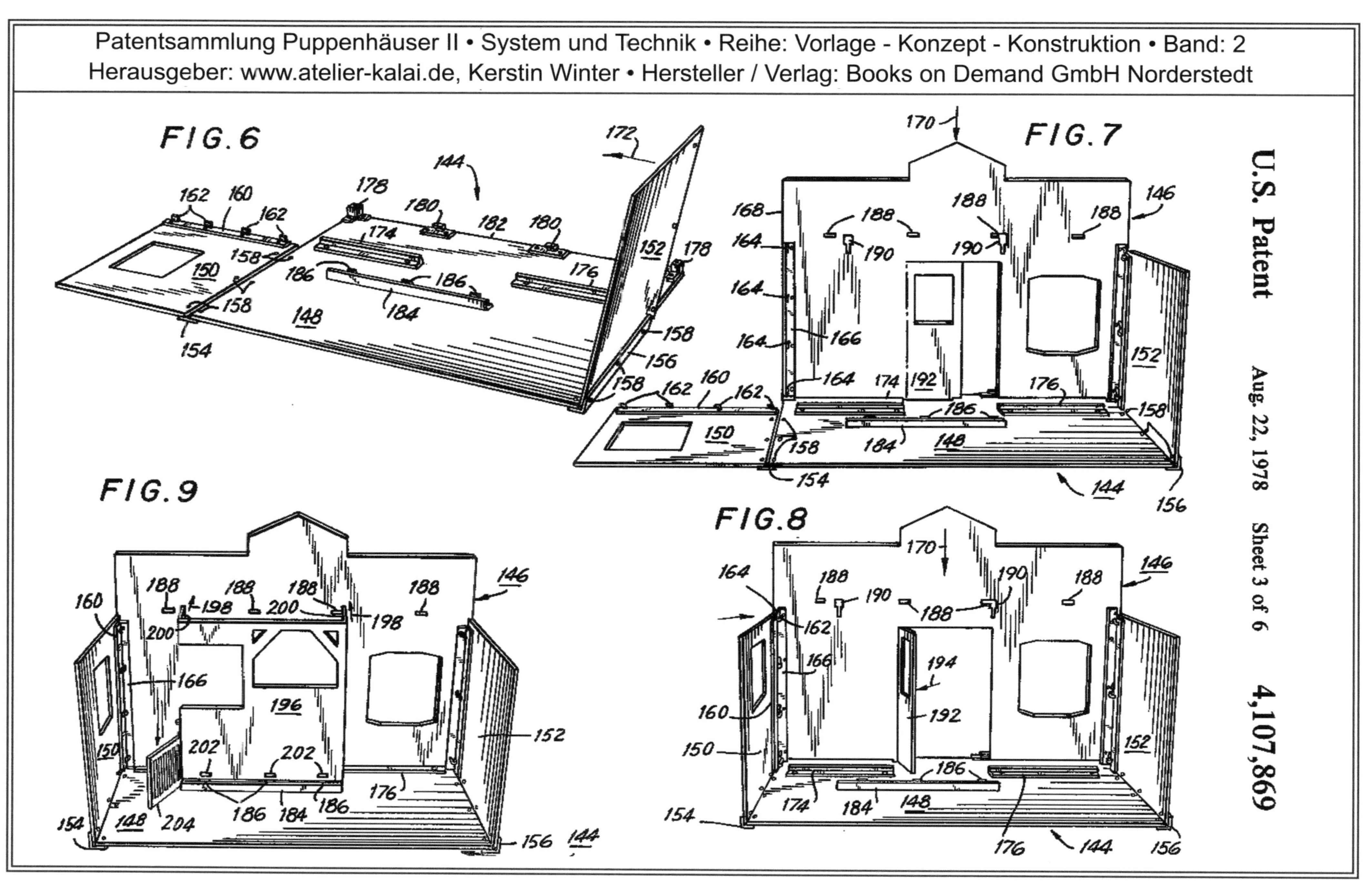

U.S. Patent
Aug. 22, 1978
Sheet 3 of 6
4,107,869
FIG.6
FIG.7
FIG.8
FIG.9

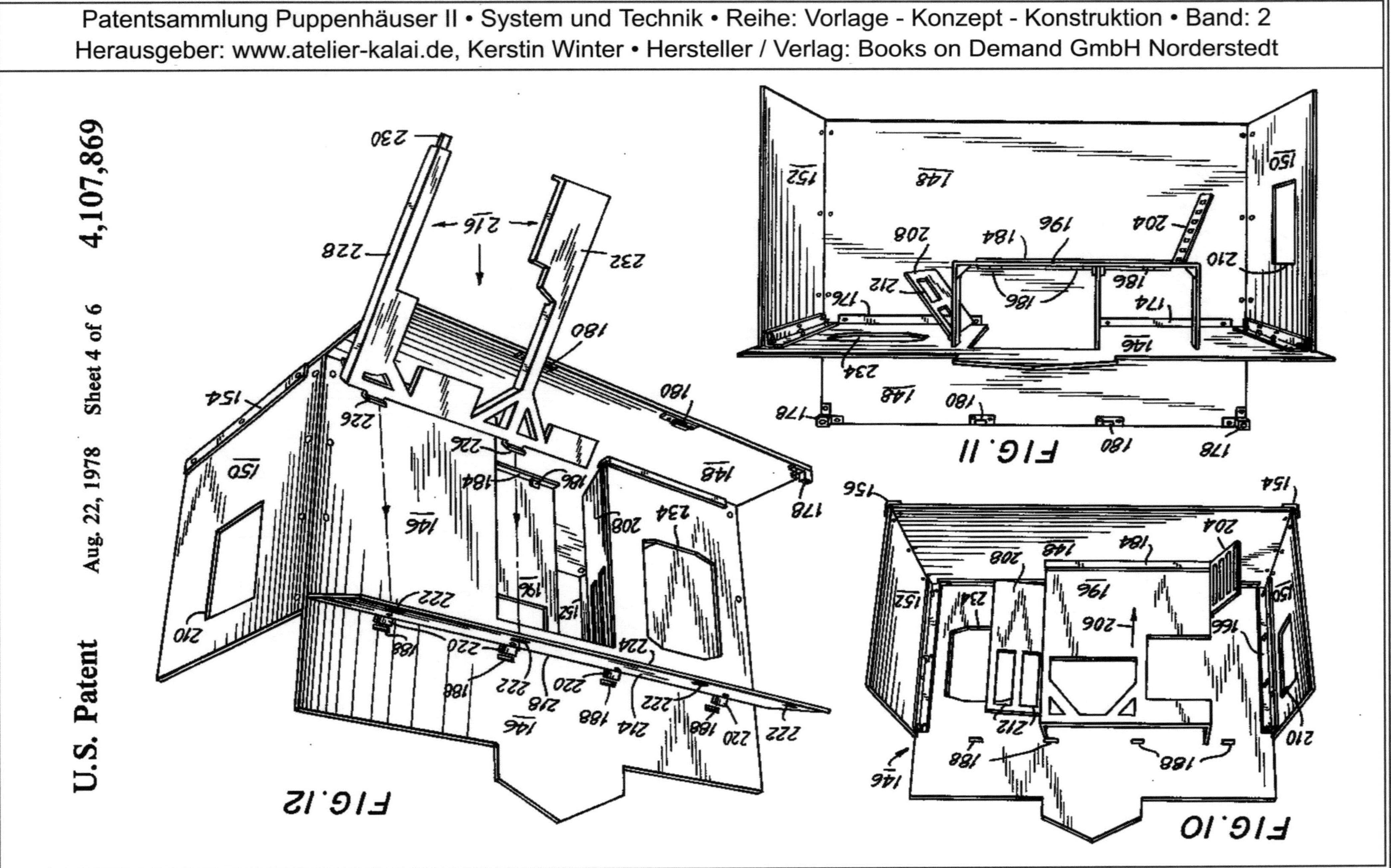

U.S. Patent
Aug. 22, 1978
Sheet 4 of 6
4,107,869
FIG.12
FIG.11
FIG.10

FIG.13

FIG.14

FIG.15

FIG.16

FIG.17

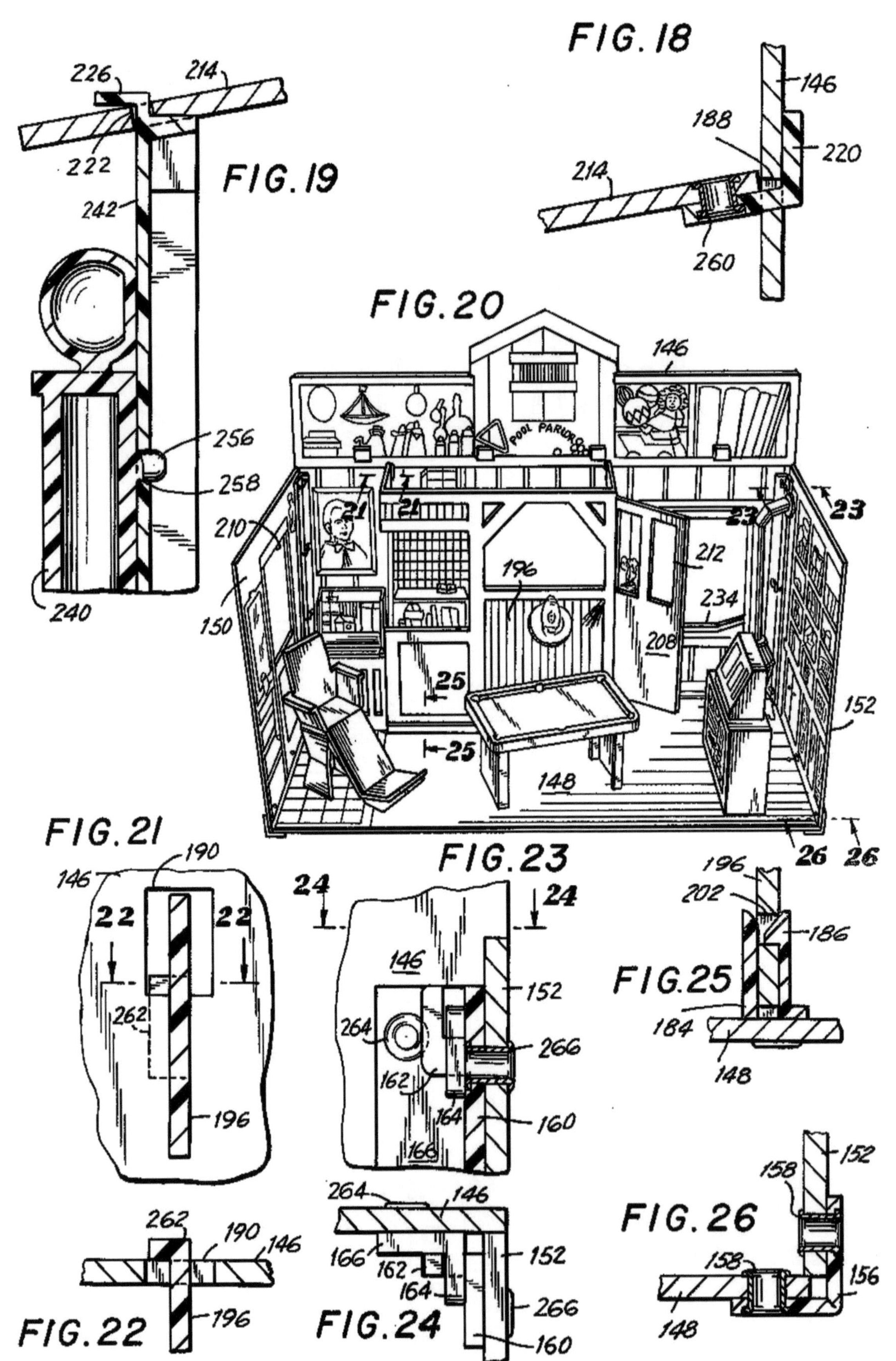

Patentsammlung Puppenhäuser II • System und Technik • Reihe: Vorlage - Konzept - Konstruktion • Band: 2
Herausgeber: www.atelier-kalai.de, Kerstin Winter • Hersteller / Verlag: Books on Demand GmbH Norderstedt

1

DEMOUNTABLE TOY HOUSE

BACKGROUND OF THE INVENTION

1. Field of the Invention

A toy house.

2. Description of the Prior Art

Toy houses provide much enjoyment for children, because of the simulation of ownership of real property and the stimulation of being in exclusive possession of territory, namely the interior of the toy house. This is especially true when small scale appurtenances such as toy furnishings and toy figures are also provided, to fit inside the toy house.

Among the abundance of recent prior art related to this field of art may be mentioned the following U.S. patents which relate to three-dimensional toy houses: U.S. Pat. Nos. 3,906,659; 3,888,039; 3,872,620; 3,849,930; 3,751,848; 3,729,881; 3,719,001; 3,629,969; 3,597,858; 3,548,552; 3,446,790 and 3,363,360. In general, the prior art relies on separate pins and/or clips or the like to hold the several panels of a toy house together and thus to provide a unitary structure.

SUMMARY OF THE INVENTION

1. Purposes of the Invention

It is an object of the present invention to provide an improved toy house.

Another object is to provide a toy house which may be assembled without the usage of separate pins, clips, bolts or the like.

A further object is to provide a toy house which is simply and easily assembled by a child.

An additional object is to provide a toy house which is of low cost to manufacture while providing the appearance of a real house.

Still another object is to provide a toy house which is rugged and durable when assembled, and which is not easily broken by a child at play.

These and other objects and advantages of the present invention will become evident from the description which follows.

2. Brief Description of the Invention

In the present invention, at least one vertically oriented interior panel is provided in the toy house. In addition, at least one inclined roof panel and a horizontal flat floor panel are provided. The floor panel is provided with integral means to provide restraint of the first panel so as to maintain it in an upright position. In addition, the roof panel is supported in an inclined position by means which are also restrained by the integral means in the floor panel.

In one embodiment of the invention, a pair of perpendicular vertical panels are provided in the interior of the toy house, which is assembled from a plurality of flat planar panels. A first vertical interior panel has an upper vertical slot, two upper edges which slope downwardly away from the open upper end of the slot, two vertical side edges, each of which is defined by a vertical strip having two lips which extended transversely to the side edges of the panel, and a lower horizontal edge having a terminal vertical extension at each end. A second vertical interior panel is provided. The second panel is oriented generally perpendicular to the first panel, and is a generally rectangular panel. The second panel is provided with a lower vertical slot which extends upwards from the lower horizontal edge of the second panel, so that each of the first and second panels fits into

2

the respective slot of the other panel and thus the two panels are oriented in a disposition perpendicular to each other. The lower horizontal edge of the second panel is provided with terminal vertically downward extensions at each end thereof. The second panel also has two vertical side edges, each of which is defined by a vertical strip. Each vertical strip of the second panel has an upper recess and two lips which extend transversely to a side edge of the second panel. Two generally rectangular roof panels are also provided. Each roof panel is provided with a horizontal lip which extends transversely to the top edge of the panel. Each of these lips projects outwards from the corners of the panel, so that when the toy house is assembled, each lip extends into one of the aforementioned upper recesses in a vertical strip of the second panel. Finally, a fifth panel is provided as a floor panel of the toy house. This floor panel is a horizontally oriented generally rectangular panel, with a slot extending inwards from the periphery of each edge of the floor panel. When the toy house is assembled, each terminal vertically downward extension of the first and second panels extends into one of the slots in the floor panel. Thus the toy house is assembled into a rigid configuration without the use of bolts, pins or clips.

In preferred embodiments of this configuration of the invention, the toy house may be provided with a rectangular parallelepiped chimney. Two opposed lower edges of the chimney conform to the dimension of the toy house, i.e. each of two opposed lower edges of the chimney slope upwards from either end to a central notch. The central notches in the opposed lower edges of the chimney fit over the juxtaposed lips of the roof panels when the toy house is assembled. The outer corners of the roof panels and the corners of the floor panel will preferably be curved, i.e. rounded. Typically the toy house in this embodiment described supra is a wooden house, i.e. the panels are composed of wood, e.g. plywood. A rectangular horizontal panel simulating a patio may also be provided. One edge of this panel is provided with a notch, which notch fits over the lower end of an edge strip of the first or second panels when the toy house is assembled. The outer corners of the patio panel may be rounded. Finally, in a preferred embodiment, the several members are symmetrically oriented, i.e. the upper vertical slot of the first vertical panel and the lower vertical slot of the second vertical panel are each centrally located in the respective panel, and the slots in the floor panel are each located generally about at the center of the respective side edge of the floor panel.

In an alternative embodiment of the invention, at least one of the panels includes a flexible and resilient bendable connection between sections. Typically, in this embodiment, a first panel consisting of a vertically oriented interior panel is provided. A second panel which is horizontally oriented for assembly is also provided. The second panel includes a central floor section and at least one lateral section which is connected along one edge to the central floor section by a bendable flexible and resilient connection, so that the lateral section is capable of being elevated to form an upright wall section perpendicular to the central floor section. Connection means are provided along one side edge of the upright wall section, and mating connection means are provided along a vertical edge of the first panel, so that the first panel and the upright wall section of the second panel are connectable perpendicular to each other. One

4,107,869

3

or more roof panels are provided, with a roof panel being mountable along one edge to the first panel and supportable along its opposite edge by a plurality of stanchions or the like. Each stanchion extends generally vertically, from attachment to the roof panel, to attachment to mounting means provided along an edge of the central floor section of the second panel.

In a preferred embodiment of this alternative embodiment, the second panel has two lateral sections on opposite sides of the central floor section. Each lateral section is convertible into an upright vertical wall section, perpendicular to the central horizontal floor section, through the provision of a bendable connection between each lateral section and the central floor section. A plurality of connection means are provided along both vertical edges of the first panel, each of which connection means is matable with connection means along one side edge of each of the upright wall sections, so that the first panel and each of the upright wall sections are perpendicularly connectable, with the upright wall sections being parallel and in registration when both upright wall sections are connected to the first panel. The matable connection means for perpendicular connection of the first panel and each of the upright wall sections will typically consist of a plurality of pins on one member, each of which is slidable into lock position in a corresponding recess in the other member.

The third or roof panel is preferably mounted along one edge to the first panel by providing a plurality of spaced apart tabs along the edge of the third panel, and a corresponding plurality of tab entry openings in the first panel. Usually and preferably, each stanchion is provided with an upper tab and a lower protuberance. The upper tabs fit into openings in the opposite edge of the third (roof) panel, and the lower protuberances fit into recesses along an edge of the central floor section of the second panel.

The lower edge of the first panel will usually be contiguous with the central floor section of the second panel when this alternative embodiment of the toy house is assembled, and the second panel preferably extends outwards on both sides of the contiguous junction with the first panel. The reason for this arrangement is so that one side of the assemblage may simulate the exterior, e.g. front porch, of the toy house; while the other side of the assemblage simulates the interior of the toy house. In this case, the first panel will usually be provided with an integral door or means simulating a door. Also, the third panel (roof) will be mounted above one side of the second panel, to simulate a front porch of the toy house, and the other side of the second panel will be provided with means simulating the interior of the toy house, such as a toy stove, toy furniture, etc., mounted on the central floor section of the second panel, and appropriate legends, illustrations and/or pictures will be provided on either side of the first panel and also on the upright wall sections of the second panel. Thus the panels will typically be composed of laminated cardboard having plastic edges and joining pieces, with the laminated cardboard panels being flat pieces of laminated paper, each panel being printed over legends and/or pictures, and each panel being covered on both sides with a thin plastic film composed of a plastic material such as polyethylene, polyvinyl chloride, mylar, cellulose acetate, cellulose acetate butyrate, nylon or an acrylic resin such as methyl methacrylate or lucite.

4

The toy house configuration of the present invention provides several salient advantages. The toy house is assembled without the usage of pins, clips, bolts or the like and thus small parts are not employed to hold the panels together. Therefore small parts such as pins or clips cannot become lost during handling or assembly, which loss could preclude proper assemblage of the toy house. The present toy house is simple in structural configuration and thus is easily assembled by a child. The toy house of the present invention is of low cost of manufacture while still providing the appearance and illusion of a real house, thus providing children with an inexpensive and attractive toy. The present toy house is rugged and durable when assembled, and is not easily broken by a child at play.

The invention accordingly consists in the features of construction, combination of elements, and arrangement of parts which will be exemplified in the article of manufacture hereinafter described and of which the scope of application will be indicated in the appended claims.

BRIEF DESCRIPTION OF THE DRAWINGS

In the accompanying drawings in which are shown several of the various possible embodiments of the invention:

FIG. 1 is an exploded perspective view of one embodiment of the present toy house;

FIG. 2 is a perspective view of the assembled toy house of FIG. 1;

FIG. 3 is a sectional elevation view taken substantially along the line 3—3 of FIG. 2;

FIG. 4 is a sectional elevation view taken substantially along the line 4—4 of FIG. 2;

FIG. 5 is a partial sectional plan view taken substantially along the line 5—5 of FIG. 3;

FIG. 6 is a perspective view of a panel member of an alternative embodiment of the invention;

FIG. 7 shows the coaction and assembly steps of two panels of the alternative embodiment;

FIGS. 8, 9 and 10 show further assembly steps of the alternative embodiment of the invention;

FIG. 11 is a perspective plan view of the partially erected alternative embodiment;

FIGS. 12 and 13 show emplacement of the roof panel and stanchions in the alternative embodiment;

FIG. 14 shows the completed front of the toy house, in this case having the appurtenances of a toy country store;

FIGS. 15 and 16 are partial sectional elevation views taken substantially along the lines 15—15 and 16—16 of FIG. 14 and showing stanchion joints;

FIG. 17 is a partial sectional plan view taken substantially along the line 17—17 of FIG. 14 and showing a stanchion joint;

FIGS. 18 and 19 are partial sectional elevation views taken substantially along the lines 18—18 and 19—19 of FIG. 14 and showing further details of joints;

FIG. 20 is a perspective view of the rear or interior of the country store of FIG. 14;

FIG. 21 is a sectional elevation view of a joint taken substantially along the line 21—21 of FIG. 20;

FIG. 22 is a sectional plan view of the joint of FIG. 21, taken substantially along the line 22—22;

FIG. 23 is a sectional elevation view of another joint, taken substantially along the line 23—23 of FIG. 20;

FIG. 24 is a sectional plan view of the joint of FIG. 23, taken substantially along the line 24—24; and

5

FIGS. 25 and 26 are sectional elevation views of still further joints in FIG. 20, taken substantially along the lines 25—25 and 26—26 of FIG. 20.

DETAILED DESCRIPTION OF THE PREFERRED EMBODIMENTS

Referring now to FIGS. 1–5 and especially to FIG. 1, the the several members of the toy house are shown separated from each other and will be described in detail. The toy house is characterized by the provision of a plurality of interconnected flat planar panels, as shown in FIG. 1. These panels include a first interior panel 30. The panel 30 is one of the two vertically oriented interior panels disposed at right angles, i.e. perpendicular to each other, in this embodiment of the invention. The panel 30 has an upper vertical slot 32, two upper edges 34, 36 which slope downwardly away from the open upper end of slot 32, and two vertical side edges defined by vertical strips 38, 40, each vertical strip having two lips which extend transversely to a side edge of the panel. Thus strip 38 includes lips 42, 44 and strip 40 includes lips 46, 48. The panel 30 is completed by the provision of a lower horizontal edge 50 having terminal vertical downward extensions 52, 54 at its end.

A second vertically oriented interior panel 56 is disposed at a right angle to the panel 30, i.e. the panels 30 and 56 are perpendicular to each other when the toy house is assembled. The panel 56 is a generally rectangular panel having a lower vertical slot 58, which slot 58 extends upwards from the lower horizontal edge 60 of the panel 56. Thus when the toy house is assembled, the panels 30 and 56 are generally perpendicular to each other, with the upper central portion of panel 56 fitting into slot 32 and the lower central portion of panel 30 fitting into slot 58. The lower edge 60 of the panel 56 has terminal vertical downward extensions 62 and 64 at its ends. The two vertical side edges of panel 56 are each defined by a vertical strip 66, 68. The vertical strip 66 has an upper recess 70 and two lips 72, 74 which extend transversely to the side edge of the panel 56. Similarly, the vertical strip 68 has an upper recess 76 and two lips 78, 80 which extend transversely to the opposite side edge of the panel 56.

The toy house is also provided with a third and fourth panels consisting of generally rectangular roof panels 82 and 84. The roof panel 82 has a horizontal lip 86 which extends transversely to the top edge of panel 82, and roof panel 84 has a similar horizontal lip 88 which extends transversely to the top edge of panel 84. The ends of each lip 86, 88 projects outward from the corners of the respective panel 82, 84. Thus lip 86 has terminal extensions or projections 90, 92 and lip 88 has terminal extensions or projections or projections 94, 96. As will appear infra, when the toy house is assembled, each terminal extension 90, 92, 94 or 96 extends into one of the upper recesses 70, 76 of a respective vertical strip 66, 68. The outer corners 98, 100 of panel 82 and the outer corners 102, 104 of panel 84 are preferably curved, i.e. rounded as shown, to provide a pleasing effect and to prevent injury to children playing with the toy house.

The toy house is completed in its broadest and most general embodiment by the provision of a fifth panel 106 which is a generally rectangular horizontal floor panel. The panel 106 is characterized by the provision of slots 108, 110, 112 and 114. Each of these slots extends inwards from an edge of the periphery of the panel 106, so that each terminal vertically downward

6

extension of the lower edges of the panels 30 and 56 fits into one of the slots when the toy house is assembled, to provide structural rigidity to the toy house. Thus, extension 52 fits into slot 108, extension 54 fits into slot 112, extension 62 fits into slot 114 and extension 64 fits into slot 110, when the toy house is assembled.

Various preferred embodiments and appurtenances of the toy house are shown and will be described. At the onset, it is preferred for reasons of symmetry and stability that the vertical slots 32 and 58 be centrally located in the respective panels 30 and 56, and that the slots 108, 110, 112 and 114 are each located at about the center of the respective side edge of the rectangular floor panel 106. These configurations of the slots lead to a symmetrical toy house, which is desired for reasons of appearance, strength and stability.

In addition, two optional appurtenances for the toy house are shown in FIG. 1, namely a chimney 116 and a horizontal patio member 118. The chimney 116 is of rectangular parallelepiped configuration. The lower edge of a front panel 120 of the chimney 116 is of a specific configuration to accommodate emplacement of the chimney anywhere along the peak of the toy house as desired. Thus, the front panel 120 of chimney 116 has a lower edge having two portions 122, 124 which slope upwards from the ends, i.e. corners, of the chimney to a central notch 126. The opposed lower edge of the opposite back panel 128 of chimney 116 is of a similar configuration, so that the notches such as 126 fit over the juxtaposed lips 86, 88 of the respective roof panels 82, 84 when the toy house is assembled, and also the sloping edge portions 122, 124 fit nicely onto the respective roof panel surfaces 82, 84, which panel surfaces in turn are sloped because of the slope in edges 34, 36.

The patio panel 118 is an optional additional sixth panel in the toy house combination. The panel 118 is a generally rectangular horizontal patio panel, with a notch 130 being provided in one edge of the panel, which notch 130 fits over the lower end of the strip 68 when the toy house is assembled. It is evident that the patio panel 118 may be disposed adjacent to any side of the toy house, thus notch 130 could alternatively accommodate for the lower end of strips 66, 38 or 40.

As was the case with corners 98, 100, 102 and 104, it is evident that corners 132, 134, 136 and 138 of panel 106 and corners 140 and 142 of panel 118 are preferably curved, i.e. rounded. In addition, it is apparent that more than one chimney 116 and/or more than one patio panel 118 may be provided in practice.

FIGS. 2, 3, 4 and 5 illustrate the fully assembled toy house. The coaction between the panels and appurtenances thereto is as described supra, thus the vertical panels 30 and 56 are at right angles with the central vertical slot in each panel accommodating the other panel. The lower vertically downward extensions 52, 54 in panel 30 and 62, 64 in panel 56 each fit into one of the slots in horizontal floor panel 106. FIG. 5 illustrates this arrangement in detail, as well as the fitting of the patio panel 118 into the assemblage. Miniature or toy articles of furniture are shown in phantom outline in FIGS. 2, 3 and 4, with each quarter section of the toy house serving as a simulation of a different room of the house, e.g. living room, kitchen, bedroom.

Referring now to FIGS. 6, 7 and 8, one panel member 144 of an alternative embodiment of the invention is shown. The panel 144 for purposes of reference will be referred to as the second panel of the combination, with the first panel being shown in FIG. 7 as a vertically

7

oriented interior panel 146. The panel 144, which is the second panel in the combination, is horizontally oriented for assembly and includes a central floor section 148 and lateral sections 150 and 152. Each lateral section 150, 152 is connected along one edge to the central floor section 148 by a bendable, i.e. flexible and resilient, connection, so that the lateral sections, note section 152 in FIG. 6, are capable of being elevated to form upright wall sections perpendicular to the central floor section 148. The bendable connections 154 between sections 148 and 150, and 156 between sections 148 and 152, consist generally of strips of any suitable flexible and resilient material, e.g. a plastic such as polyethylene, polypropylene, especially isotactic polypropylene, or polyvinyl chloride, natural or synthetic rubber, etc.

Bendable connections 154 and 156 are attached to the respective members 148 and 150 or 152 by means of suitable rivets, bolts, or staples, generally designated as 158. The attachments 158 are permanent attachments, and in order to allow the connections 154 and 156 to bend, a longitudinal groove may be provided in each of these members. The panels 144 and 146 may be composed of any suitable material e.g. a plastic similar to that used for the strips 154 and 156, however the panels 144 and 146 are preferably composed of laminated cardboard, with the outer surfaces of the panels as well as the joining pieces being composed of plastic. Thus basically the panels 144 and 146 in most instances will be laminated cardboard panels which essentially consist of flat pieces of laminated paper, with each panel being printed with legends and/or pictures, and with each panel being covered on both sides with a plastic film. This plastic film consists generally of transparent plastic material such as polyethylene, polyvinyl chloride, mylar, cellulose acetate, cellulose acetate butyrate, nylon, or acrylic resin such as methyl methacrylate or lucite.

As best shown in FIG. 6, panel member 150, which when assembled into the toy house will be an upright wall section perpendicular to the central floor section 148, is provided with a plastic stiffening rib 160 along one edge, which rib 160 will be vertically oriented when the toy house is assembled. The rib 160 is provided with a plurality of spaced apart L-shaped protrusions 162, which as will appear infra, are slidable into locked position with corresponding spaced apart L-shaped protrusions along a vertical edge of the panel 146. This cooperation between L-shaped protrusions is best shown in FIGS. 7, 8 and 9. In these figures the interlocking between protrusions 162 and a corresponding plurality of protrusions 164 on rib 166 disposed along a vertical edge 168 of panel 146 is evident. To attain this cooperation, the outer arm of each protrusion 162 extends upwards, while the outer arm of each protrusion 164 extends downwards, so that the protrusions 162 and 164 are slidable together into lock position. A similar connection means is provided between the upright wall section 152 of panel 144, and panel 146. Referring to FIG. 7, the arrow 170 shows the direction of manual manipulation of panel 146 vertically downwards to accomplish the emplacement of panel 146 in conjunction with the maintenance of panels 150 and 152 in an upright position, which is also manually accomplished as shown by arrow 172 (FIG. 6). Stop members 174 and 176 are provided in order to maintain panel 146 in position.

FIG. 6 also shows other elements which are provided for the coupling of members of the toy house to panel 144. Thus FIG. 6 shows recesses 178 at adjacent corners

8

of the central floor section 148 of the second panel 144, as well as raised flat horizontal ribs 180 along edge 182 of section 148. The members 178, 180 are mounted to section 148 by an appropriate means comparable to ribs 160 and 166 together with fastening means such as elements 158 described supra. Finally, FIG. 6 shows a mounting element 184 which is provided with a plurality of clip members 186, so that an appurtenance of the toy house may be clipped into place. Referring now to FIG. 7, the panel 146 is provided with a plurality of spaced apart horizontal slots 188, as well as two openings 190 having a wide upper portion and a narrower lower portion. Referring now to FIGS. 7 and 8, a simulated door 192 is provided, and this door 192 opens inwards as shown by the arrow 194 (FIG. 8). Referring now to FIG. 9 and 10, the emplacement of a partition member 196 which simulates an interior partition in a country store is shown. FIG. 9 shows (note arrows 198) how the member 196 is disposed for emplacement. The terminal vertical edges 200 of member 196 are each provided with a protruding rib, and the ribs are inserted through the wide portion of the respective opening 190; thereafter the member 196 is moved downwards so that the rib in each case slidingly engages the narrow lower portion of the opening 190. Concomitantly, the lower slots 202 of member 196 engage members 186, as will appear infra. FIG. 9 also shows the emplacement of a gate 204 along one side edge of partition member 196.

FIG. 10 shows the fully assembled simulated interior of the toy house, which in this case is a toy store e.g. a toy country store as will appear infra. The final stage in the assembly is indicated by arrow 206 in which member 196 is moved downwards into position with its lower edge held by member 184. In addition a simulated door 208 has been emplaced on the side of member 196. Appropriate spacings or openings are provided to simulate a real store, such as window opening 210 in panel member 150 and openings 212 in door member 208. FIG. 11 illustrates how the front of the toy house, which is provided with members 178 and 180, extends outwards as a porch from the front of panel 146. Thus in FIG. 11, the lower porch of central floor section 148 provides the floor of the interior of the toy house while the upright wall sections 150 and 152 provide the walls of the toy house.

FIG. 12 illustrates the initial stages of assembly of the front porch of the toy country store. The principal members of the front porch assembly are a third i.e. roof, panel 214, and a bifurcated stanchion 216, which in this embodiment of the invention is one of two mirror-image stanchions. It will be apparent that the provision of a bifurcated stanchion of the specific configuration of stanchion 216 is a preferred embodiment of the invention, and that other suitable stanchion configurations may be adopted in practice.

Referring first to the roof panel 214, this third main panel of the toy house is mounted along an edge 218 to the first panel 146 by providing a plurality of spaced apart L-shaped tabs 220 along the edge of the third panel. The L shape of the tabs 220 is primarily for reasons of structural integrity, and it will be understood that other suitable tab shapes may be provided in practice. It therefore will be understood that the term "L-Shaped" refers primarily to the fact that the tab end is generally perpendicular to the surface of panel 214. A plurality of tab entry openings described supra as openings 188 are provided in the first panel 146, so that the rear edge 218 of panel 214 may readily be emplaced by

9

disposing panel 214 at an acute angle relative to panel 146 as shown in FIG. 12 and inserting each tab 220 into the respective slot opening 188. The panel 214 is also provided with a plurality of slots 222 along the edge 224 opposite to edge 218. These slots 222 accommodate the upper L-shaped tabs 226 in the stanchion 216, so that the tabs 226 fit into and through openings 222 in a manner similar to the insertion of tabs 220 described supra. The major salient difference between the disposition of the tabs is that when the toy country store is fully assembled the ends of the tabs 226 will lie horizontally on top of panel 214, whereas in the case of tab 220, the ends of these tabs will be vertically adjacent to the rear surface of panel 146. One leg 228 of the stanchion 216 is provided with the lower protuberance 230, which as will appear infra fits into a recess 178. The other leg 232 of stanchion 216 has a lower opening to accommodate rib 180, as will appear infra. The panel 146 is also provided with a window 234 to simulate the front of a store. FIG. 13 shows the final stages of assembly of the front porch of the toy house i.e. the toy country store. In this case, the tabs 226 have been inserted into openings 222 in panel 214, and the stanchions are moving rearwards as shown by arrow 236 and downwards as shown by arrow 238, to emplace the lower ends of the stanchions in the fittings 178 and 180 provided in panel 148.

FIG. 14 shows a fully assembled front porch of the toy country store, including various ancillary items which are provided to enhance the illusion of a real country store. In addition suitable labels have been affixed to various items so that they are identified and readily related to by a child. Thus for example, a simulated barber pole 240 has been mounted on a leg 242 of the second bifurcated stanchion. FIGS. 15, 16, 17, 18, and 19 show sectional views illustrating details of the assembly joints. Thus FIG. 15 shows the protuberance 230 extending into a recess member 178. In addition, a typical metallic rivet 244 used for permanent attachment of member 178, which is composed of a plastic, to panel 148, which is composed of cardboard, is shown. FIGS. 16 and 17 show details of the joint between stanchion leg 232 and panel 148. Thus rib 180 has been inserted upwards through the wider section 246 of an opening in the base of leg 232, thereafter leg 232 has been moved laterally from its initial position in which rib 180 is shown in phantom outline, to its final position in which rib 180 extends above narrow slot section 248 of the lower opening in the base of leg 232. Attachment rivets 250, which are similar in configuration and function to rivet 244 described supra are also shown in phantom outline. These rivets 250 serve to hold the lower base portion 252 of rib 180 in position, with rib 180 being joined to base 252 by the narrow neck 254.

FIG. 18 shows the disposition of tab 220 in a slot 188, as well as another rivet 260 which is similar in configuration and function to rivet 244 described supra in that rivet 260 attaches the L-shaped plastic tab 220 to the cardboard panel 214.

FIG. 19 shows how tab 226 is disposed in the opening of slot 222. FIG. 19 also shows how the simulated barber pole 240 is mounted along stanchion leg 242 by horizontal protuberance 256 which extends through a hole 258 in stanchion leg 242.

FIG. 20 shows the interior of the toy house consisting in this case of a simulated toy country store. Labels have been affixed to the panels and small plastic simulated items of furnishings are shown within the interior

10

of the toy country store. FIGS. 21 and 22 show the joint between member 196 and panel 146 including a lip 262 which extends laterally from the edge of the side panel of member 196. FIGS. 23 and 24 show a typical joint between the upright side wall section 152 which consists of a lateral section of panel 148 which has been converted into an upright vertical wall section as described supra; and the panel 146. This joint is essentially between L-shaped or hook-shaped elements 162 and 164, which interlock is shown. Rivets 264 and 266 which join the plastic mountings 166 and 160 to the respective cardboard panels 146 and 152 are also shown. FIG. 25 shows details of the joint between the base of member 196 and floor panel 148. The retention in this case is attained by the lip of element 186 extending to slot 202. FIG. 26 shows in detail the configuration of the bendable, i.e. flexible and resilient, connection 156 between members 148 and 152 in the fully assembled toy house, including rivets 158.

It thus will be seen that there is provided a toy house which achieves the various objects of the invention and which is well adapted to meet the conditions of practical use.

As various possible embodiments might be made of the above invention, and as various changes might be made in the embodiments above set forth, it is to be understood that all matter herein described or shown in the accompanying drawings is to be interpreted as illustrative and not in a limiting sense.

Having thus described the invention, there is claimed as new and desired to be secured by Letters Patent:

1. A toy house assembled from a plurality of panels and comprising a plurality of flat planar panels, said panels including

(a) a first panel, said first panel being a vertically oriented interior panel having an upper vertical slot, two upper edges which slope downwardly away from the open upper end of said upper slot, two vertical side edges, each of said side edges being defined by a vertical strip, each vertical strip having two lips which extend transversely to a side edge of said first panel, and a lower horizontal edge having a terminal vertical extension at each end thereof;

(b) a second panel, said second panel being a generally rectangular vertically oriented interior panel having a lower vertical slot, said lower slot extending upwards from the lower horizontal edge of said second panel, so that when the toy house is assembled, said first and second panels are substantially perpendicular to each other with each of said first and second panels fitting into the respective slot of the other panel, the lower horizontal edge of said second panel having a terminal vertical extension at each end thereof, and two vertical side edges, each of said side edges being defined by a vertical strip, each vertical strip having an upper recess and two lips which extend transversely to a side edge of said second panel;

(c) a third and fourth panels, said third and fourth panels being generally rectangular roof panels, each of said third and fourth panels having a horizontal lip which extends transversely to the top edge of the respective associated panel, each lip projecting outwards from a corner of said associated panel so that each lip extends into one of said upper recesses in a vertical strip of a said second panel when the toy house is assembled; and

11

(d) a fifth panel, said fifth panel being a generally rectangular horizontal floor panel, each edge of said fifth panel having a slot which extends inwards from the periphery of said fifth panel, so that each terminal vertical extension of said first and second panels extends into a slot when the toy house is assembled.

2. The toy house of claim 1 together with a rectangular parallelepiped chimney, each of two opposed lower edges of said chimney sloping upwards from either end to a central notch, said notches fitting over the juxtaposed lips of the third and fourth panels when the toy house is assembled.

3. The toy house of claim 1 in which the outer corners of the third and fourth panels and the corners of the fifth panel are rounded.

12

4. The toy house of claim 1 in which the panels are composed of wood.

5. The toy house of claim 4 in which the wood panels are plywood.

6. The toy house of claim 1 together with a sixth panel, said sixth panel being a generally rectangular horizontal patio panel, one edge of said sixth panel having a notch, said notch fitting over the lower end of a strip of the first or second panel when the toy house is assembled.

7. The toy house of claim 6 in which the outer corners of the sixth panel are rounded.

8. The toy house of claim 1 in which the upper vertical slot of the first panel and the lower vertical slot of the second panel are each centrally located in the respective panel, and the slots in the fifth panel are each located about at the center of the respective side edge of the fifth panel.

* * * * *

(No Model.)

W. STRANDERS.
BUILDING TOY.

No. 311,793.

Patented Feb. 3, 1885.

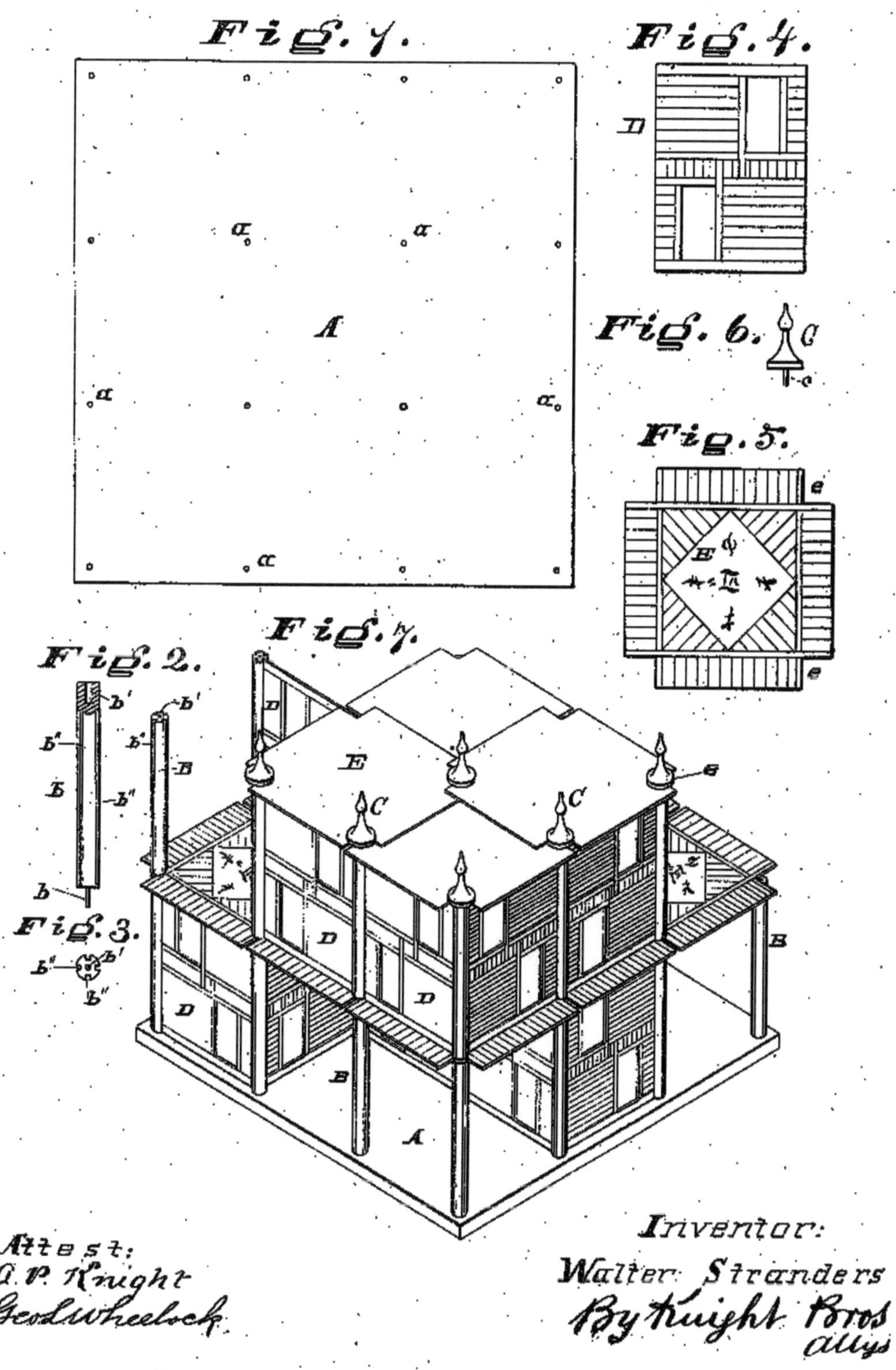

Attest:
A. P. Knight
Geo. L. Wheelock

Inventor:
Walter Stranders
By Knight Bros
Attys

UNITED STATES PATENT OFFICE.

WALTER STRANDERS, OF NEW YORK, N. Y., ASSIGNOR TO PETER G. THOMSON, OF CINCINNATI, OHIO.

BUILDING-TOY.

SPECIFICATION forming part of Letters Patent No. 311,793, dated February 3, 1885.

Application filed July 18, 1884. (No model.)

To all whom it may concern:

Be it known that I, WALTER STRANDERS, of the city of New York, in the county and State of New York, have invented a new and useful Building-Toy, of which the following is a specification.

My toy consists, essentially, of a base board which has several equidistant pits or holes to receive the doweled ends of posts or columns, the opposite end of each of which has a corresponding socket for a similar dowel, and of which each post has four equidistant longitudinal grooves to receive partitions that are carved or painted to simulate the details of a house-front, such as weather-boarding, doors, windows, &c. Other boards are provided to represent floors, roof-pieces, &c., and any desired number of additional posts, partitions, &c., to represent additional flats, stories, porches, &c. Doweled finials inserted in the post-sockets of the (for the time being) upper most tier serve to hold the roof-pieces in place, and also to impart an attractive finish.

In the accompanying drawings, Figure 1 is a plan of a base-board such as used by me. Figs. 2 and 3 are respectively a side view and an end view of a post. Fig. 4 represents a partition. Fig. 5 represents a floor or roof piece. Fig. 6 represents a finial. Fig. 7 represents one of a great variety of pleasing and instructive structures capable of being produced by different arrangements of the few simple pieces mentioned.

A represents a wooden board, which may have the square form here shown, and which has arranged in equidistant lines several holes or pits, *a*, to receive dowels *b* of posts B. Each post has a corresponding orifice or socket, *b'*, in its other end, to receive a like dowel projection either from a similar post or from one of a number of finials, C. Each post has four equidistant longitudinal grooves, *b''*, to receive the edges of wall-pieces or partitions D.

E represents floor or roof pieces, which, at their indented angles *e*, rest on posts B aforesaid, and are secured in place either by superimposed posts or by said finials C, as the case may be. Said finials have for this purpose dowels *c*, that, being inserted in the notches *e* of said pieces E, fit tightly in any one of the post-sockets.

I claim as new and of my invention—

The building-toy consisting of a base-board, A, having the several holes or pits *a*, a corresponding set of posts, B, having at one end a dowel, *b*, to occupy any hole *a* in said board, and in its other end a corresponding socket, *b'*, and in its sides four equidistant longitudinal grooves, *b''*, for partitions D, the same being combined with a series of notched floor and roof pieces, E, and of doweled finials C, substantially as set forth.

In testimony of which invention I hereunto set my hand.

WALTER STRANDERS.

Attest:
GEO. T. PINCKNEY,
WILLIAM G. MOTT.